Springer
Berlin
Heidelberg
New York
Barcelona
Budapest
Hong Kong
London
Milan
Paris
Santa Clara
Singapore
Tokyo

Richard Condit

Tropical Forest Census Plots:

Methods and Results from Barro Colorado Island, Panama and a Comparison with Other Plots

Springer

Richard Condit, Ph.D.
Smithsonian Tropical Research Institute
Panama

ISBN: 3-540-64144-0 Springer-Verlag Berlin Heidelberg New York
Environmental Intelligence Unit

Library of Congress Cataloging-in-Publication data

Condit, Richard, 1956-
Tropical forest census plots: methods and results from Barro Colorado Island, Panama and a comparison with other plots / Richard Condit.
p. cm. -- (Environmental intelligence unit)
Includes bibliographical references and index.
ISBN 1-57059-544-5 (alk. paper)
1. Forest surveys—Panama—Barro Colorado Island. 2. Forest surveys—Panama—Barro Colorado Island—Methodology. 3. Forest surveys. 4. Forest surveys—Methodology. I. Title. II. Series.
SD150.P2C66 1977
634.9'285—dc21 96-52380
CIP

Typesetting: R.G. Landes Company Georgetown, TX, U.S.A.

SPIN 10669733 31/3111 - 5 4 3 2 1 0 - Printed on acid-free paper

PREFACE

I joined the 50-hectare plot research project in Panama nearly 10 years ago, in the fall of 1988, thanks to Steve Hubbell at Princeton University, who had money to pay me as a post-doc. At first I worked on the genetics of trees in the plot, but soon they needed someone to manage the entire project, so I moved to the Smithsonian Tropical Research Institute just as the 1990 plot census was finishing. That was the third complete census of 240,000 individual plants.

This was not the first large and long-term research project on wild populations that I had joined. As a graduate student, I participated in a study of elephant seals in California which included nearly 20 years of data on individually tagged animals (the oldest animal I knew in that study was a 17-year old female tagged as a pup in 1964). In both this study and the large plot study in Panama, I instantly had access to a large dataset on wild animals, on which I could do population biology, my favorite kind of biology. I did not have to wait around for data to accumulate, it was already there for me to analyze.

As I joined the Smithsonian, the large plot project was becoming even larger. By then, in 1991, two other 50-hectare tropical forest plots had been censused, one in Malaysia and one in India. Over the past seven years, I have helped expand this project further by initiating large forest plots in Cameroon, the Democratic Republic of Congo (formerly Zaire), and Ecuador. Other scientists joined our network and started similar projects in Puerto Rico, Colombia, Thailand, Sri Lanka, the Philippines, and at a second site in Malaysia. What was a large study—240,000 trees in Panama—has become immense, now including around two million trees, each tagged, measured, mapped, and identified to species.

I like this kind of project. We have lots of data, collected using standardized techniques, all made available to many different scientists. Years from now, we hope to say that we have lots of long-term records on tropical forest trees. I firmly believe that this is how forest ecology, and ecology in general, has to be done. We need long-term records on population change and demographic parameters for many communities and many species. This large-scale and long-term approach is common in other sciences, as astronomers are now digitizing full-sky maps and geneticists are sequencing complete genomes. The purpose of all these large projects is to provide truly cohesive databanks, allowing wide-scale or long-term studies that can be based on truly comparable information. The key is that the data be comparable, assembled using standardized techniques at different sites, whether referring to galactic distributions, DNA sequence, or tree growth.

This is what we have promulgated with the Center for Tropical Forest Science, our network of large forest plots. We have been able to interest a large number of scientists in setting up their own forest plots, using the techniques developed in Panama over 15 years ago. We strive to get scientists at all 12 of the large plot sites to work together to ensure comparable data.

I do not believe there is a research project in ecology that matches ours in scale and in cohesiveness. We really do have forest data from different continents that are as strictly comparable as data from different continents can be. We really can make definitive comparisons on diversity, growth rate, or many other forest features without being concerned about methodological differences, and we can do it quickly, because all the data are in one place.

This book was written to further solidify this methodological standardization, to codify the techniques we are using at 12 large forest plots on three different continents, to put them all down in one place for others to consult and, I hope, to consider in their own studies.

CONTENTS

Part 1: A Network of Large Forest Plots in the Tropics **1**

1.1. The CTFS and the Standardization of Methodology **3**
Purpose of this Book 4
Layout of the Book 5

1.2. Design and Purpose of the Large Plot Network **9**
The Purpose of a Large Forest Plot 9
The CTFS Network of Plots 10
Selecting a Site 10
Personnel 12

Part 2: Field Methods for a Large Forest Plot **15**
Introduction and Overview 17

2.1. Topography and the Mapping Grid **19**
Equipment and Supplies 19
Phase 1—The Survey 20
Phase 2—Placing the 5-m Grid 24
Phase 3—Calculating the Topography and Its Accuracy 26
Rationale for the Methods and Alternative Approaches 29
Time and Labor for Placing the Grid 34

2.2. Censusing Trees **37**
Personnel 37
Equipment and Supplies 38
Plants to Include 38
Work Sequence 39
Phase 1—Demarcating Subquadrats 39
Phase 2—Data Collection 39
Phase 3—Checking the Work 55
Rationale for the Methods and Alternative Approaches 56
Time, Labor, and Costs 62

2.3. Taxonomy **65**
Personnel 65
Phase 1—Review What Is Known 66
Phase 2—The First Two Hectares 66
Phase 3—Completing 50 Ha 68
Well-Known Flora 70
Rationale for the Methods and Alternative Approaches 70
Time and Labor 71

2.4. Checking the Field Work **73**
Phase 1—Dbh Checks 73
Phase 2—Taxonomy Checks 75
Alternative Approaches to Checking 75
Time and Labor 76

2.5. Recensus **77**
Census Interval 77
Before the Recensus 77
Personnel and Work Teams 78
Equipment 78
Plants to Include 78
Data Collection 78
Taxonomy and Checking 85
A Third Census and Beyond 86
Rationale for the Methods and Alternative Approaches 87
Time and Labor 89

2.6. Affiliated Censuses at BCI **91**
Seeds and Fruits 91
Canopy Height 92
Lianas 93

2.7. Summary and Overview **95**
Similarities and Differences 95
Labor Needs 97
Equipment Needs 97
Time 97
Total Cost for a Large Plot 97

Part 3: Database Methods for a Large Forest Plot **101**
Introduction and Overview 103

3.1. Data Entry **105**
Initial Data Entry 105
Checking the Data 111
Merging the Databases 112
Equipment, Time, and Labor 114
Programs and Boxes 117

3.2. Constructing a Database for the First Census **135**
Data Fields in the Final Database 135
Data Fields Removed from the Final Database 138
Splitting the Database in Two 138
Database Correction 140
Backing Up! 141
Field Sheets for the Next Census 141
Equipment, Time, and Labor 141
Programs and Boxes 143

3.3. Constructing a Database After the Second Census 161
Recruits 161
Original Trees 161
The Final Database 164
Third and Later Censuses 166
Final Database Structure 166
Database Correction 167
Field Sheets for the Next Census 167
Appendix 3.3. POM Problems at BCI 171

3.4. Data Availability 173
BCI Data 173
The Other Large Plots 174

Part 4: Species Distribution Maps from the 50 ha Plot at Barro Colorado Island 177
Introduction 179
Habitat Map 179
Species Maps 181

References 203

Index 207

Part 1:
A Network of Large Forest Plots in the Tropics

CHAPTER 1.1

The CTFS and the Standardization of Methodology

The Center for Tropical Forest Science was opened in 1990 at the Smithsonian Tropical Research Institute. Its primary purpose was to establish a series of long-term, large-scale forest census plots throughout the tropics. At that time, there were three such plots in existence—at Barro Colorado Island in Panama, Pasoh Forest Reserve in Malaysia, and the Mudumalai Game Reserve in India. Three more were underway, at the Huai Kha Khaeng Wildlife Sanctuary in Thailand, Lambir National Park in Sarawak, Malaysia, and the Luquillo Forest in Puerto Rico. The latter is 16 hectares (ha) in area, while all others are 50 hectares (Lambir is actually 52).

The original large plot was the one on Barro Colorado Island, or BCI, in the Panama Canal. It was set up in 1980 and censused for the first time from 1981-1983. By sampling such a large area so intensely—a total of 240,000 woody plants down to 10 mm in stem diameter were mapped and identified—the BCI plot led to unprecedented conclusions about a tropical forest community (Hubbell and Foster 1983, 1986a). First, the large sample provided the most precise account of rarity then available for tropical trees (Hubbell and Foster 1986b). Second, it provided substantial samples of individual stems for many species, allowing detailed population analyses on a species-by-species basis for a large fraction of the community (Hubbell and Foster 1992; Condit et al 1992, 1994, 1995a, 1996a,b). Very few prior demographic analyses of tropical forests had allowed species-specific precision. This success led to the creation of the next two 50 ha plots, at Pasoh in Malaysia and Mudumalai in India (Manokaran et al 1992; Sukumar et al 1992).

The BCI plot was created with the intention of examining ecological theories on the maintenance of high species diversity (Condit 1995). Research from the plot has led to important progress in this field, but recently, scientists involved with the projects began to recognize other important contributions the large plots could make. By providing species-specific demographic information, they could be used for examining models of sustainable timber harvest and estimating the value of forest products (Saw et al 1991; Condit et al 1993a,b, 1995b; Condit 1995). In addition, the thorough inventories provide precise baseline information on community composition, so that long-term changes can be documented. Finally, the plots provide a resource for other scientists doing field work on tropical trees who can benefit from a map of every individual of every species (e.g., Murawski et al 1990; Stacy et al 1996).

Tropical Forest Census Plots: Methods and Results from Barro Colorado Island, Panama and a Comparison with Other Plots, by Richard Condit.

It was these multiple uses that spurred the creation of the Center for Tropical Forest Science, or CTFS. The Center's goal is to create a network of large plots, placed in a variety of the major tropical forest realms of the world, in order to find general answers to the major research questions—both applied and basic—described above. The network has doubled in size since 1990, and there are now 12 large forest dynamics plots in the tropics (see chapter 1.2). Most of the plots are directed and operated by local research agencies or institutions, often in collaboration with American, European, or Japanese scientists. The CTFS does not do the research at most sites, but develops the collaborations that lead to standardized data collection and exchange.

The most important feature of the network of plots is the standardization. CTFS has worked hard to ensure that similar methods are used at all plots so that results can be compared easily. This has been accomplished by frequent meetings between scientists working at different sites, in particular, head scientists from most of the plots have visited Barro Colorado to observe our techniques. In addition, a number of pamphlets describing research methods have been circulated and translated into various languages. One of these was published by the Forest Research Institute of Malaysia (Manokaran et al 1990).

Purpose of this Book

Comparability across so many large data sets is a remarkable opportunity. It allows identical analyses on mortality, growth, distribution, basal area, and diversity to be carried out at all sites with remarkable ease. Computer programs which do the work at one site can immediately be applied to another, and results are completely comparable. Moreover, results can lead to general hypotheses when similar patterns are found in distinct forests. My goal here is to describe the details of a recommended, standard methodology for large plots so comparability is assured, including both field methods and database methods.

There are two ways my recommendations can be used. First, they are advice to people already working on tree censuses in the tropics: I have collected solutions to myriad problems associated with a large plot. Many of these solutions were discovered after the fact, when important errors turned up in the BCI data, and I hope other workers will benefit from our mistakes. Indeed, I try here to carefully and honestly document errors made at BCI. The second use of the book is as a complete documentation of methods used at BCI—including many details that do not make it into shorter journal articles—for others to know how we have arrived at our conclusions.

The book also illustrates some of the results of the BCI census by providing distribution maps for selected species. They are intended to show the power of the information generated by all the hard work going into a large census.

A field guide for a large plot can also be a field guide for a small plot. All of the methods I describe here are perfectly applicable to 0.1 or 1 ha plots (Dallmeier et al 1991). Some of the procedures I describe, though, are obviously aimed at problems associated with very large data sets, and it is these difficulties that I wish to emphasize. Besides plot size, this book will also be applicable to studies that deviate from other aspects of our methods. For example, many forest censuses in the tropics consider only stems ≥ 100 mm in diameter (Swaine et al 1987; Phillips and Gentry 1994; Phillips et al 1994), whereas our large plots include stems down to 10 mm.

Some censuses focus on certain species, whereas CTFS plots include everything except lianas. Other studies can thus profit by matching certain aspects of our methods, because results can then be compared to subsets of the large plot data.

Layout of the Book

Part 1 of the book is a short introductory section. It provides a brief description of the existing plot network and a chapter on the criteria used by CTFS to select plot sites. These are criteria by which new plots might be added to the network.

Parts 2 and 3 are detailed recommendations on how a large forest census in the tropics should be done, first covering field methodology (Part 2) then database methodology (Part 3). These recommendations are a distillation of various methods used over the years at BCI and other plots—they are not necessarily the same as the methods followed at any one plot! The reason for this is that creating the BCI plot has been a learning process: mistakes were made along the way, and I now know how to avoid them. Some of these mistakes did not become evident until the third or fourth census was completed and novel data analyses were carried out. So my current recommendations are most certainly *not* the same as the original methods used at BCI. They represent how I would set up a large plot if I had to do it all over again. Indeed, in the plots at Yasuní in Ecuador and Korup in Cameroon, we are trying it all over again and following the recommendations presented in this book as closely as at any other plot.

Part 4 is a selection of field maps from the BCI plot. It includes distribution maps of individual species that illustrate the main patterns we have discovered, plus topographic and habitat maps.

Acknowledgments

For the information on large plots which has been funneled through me, I have to thank all the biologists, botanists, and foresters who have participated with the network. It was Steve Hubbell and Robin Foster who first formulated the idea that studying a large piece of tropical forest was useful and who figured out how to do it at BCI. Most of the methods documented in this book trace back to Steve and Robin, and both have continued close involvement at BCI. Steve also helped start the Pasoh project in Malaysia and Robin the Yasuní project in Ecuador. The next key large plot participants were Peter Ashton and Jim LaFrankie, who started the series of plots in Asia. Without their field experience and connections with local scientists, none of the Asian plots would have been started, and Jim has been responsible for developing the methods used there. Those four—Hubbell, Foster, Ashton, and LaFrankie—built the CTFS network by developing the original methodology and extending it to new sites.

Many other scientists are responsible for the direction and management of individual plots. The third plot, at Mudumalai in India, was started by R. Sukumar in India, and he was solely responsible for adapting the BCI methodology to the dry forest there. Savitri and Nimal Gunatilleke, along with Peter Ashton, have made the Sinharaja plot in Sri Lanka one of the most carefully done of all, and provided me with detailed reports on their work. N. Manokaran managed the field work at the Pasoh plot and thus was the first to make a large plot work in a truly hyperdiverse forest. Jess Zimmerman and Robert Waide established the "big grid" in Puerto Rico, and Jill Thompson provided me with details on methods used there. Terese Hart, John Hart, and I directed the plot at Ituri in the Congo, the remotest of the large plots. In Ecuador, Renato Valencia, Katya Romeleroux, Henrik Balslev, Elizabeth Losos, Robin Foster, and I started the first large plot in Amazonia. The plot in Thailand has been directed by B. Sarayuhd, the plot in Lambir by Lee Hua Seng and T. Yamakura, and the Palanan plot by Edmundo Gumpal. The latest additions are in Cameroon, where Nick Songwe, Duncan Thomas, Elizabeth Losos, and I have just begun work, and Colombia, where Christian Samper and Constanza Rios recently initiated a plot in montane forest in the Andes. Chapter 1.2 provides more information about all the plots.

Beneath all of these scientific managers have worked literally hundreds of field biologists, those who actually collect the data. I regret that it would be impossible for me to assemble all their names and thank them properly. Perhaps in the future I will make an effort to do so. I have to thank several who have worked for many years at BCI, though: Rolando Pérez, Suzanne Lao, Salomón Aguilar, and Andres Hernández. Each contributed to this book by reviewing methods with me many times over. Also, I would like to thank Pamela Hall for many discussions with me on database methods.

The botanical skill needed for identifying hundreds of thousands of sterile tropical plants is the most important skill behind a large plot. It is a remarkable ability that few people acquire. Robin Foster was the key figure at the BCI and Ecuador plots, and K. M. Kochummen, Jim LaFrankie, Sylvester Tan, and Peter Ashton were in Asia. Ewango Ndomba, Innocent Liengola Bauma, Jean-Remy Makana, along with Terry and John Hart handled the Congo flora, Duncan Thomas and David Kenfack were responsible for the Cameroon flora, H.S. Suresh the Mudumalai flora, and the Gunatilleke's the Sinharaja flora.

Two other people have directed CTFS and held it together as a real network. One is Ira Rubinoff, Director of the Smithsonian Tropical Research Institute and one of the founders of CTFS, who has been responsible for raising money and maintaining the fundamental goals of the project. The other is Elizabeth Losos, Director of CTFS, who does most of the hard work—visiting all the sites, mollifying upset partners, raising big money, as well as directing scientific and socioeconomic research.

I also thank Rolando Pérez for preparing Figures 2.2.3, 2.2.5, and 2.2.6 in chapter 2.2, and Suzanne Lao and Karl Kaufman for the programs shown in chapters 3.1, 3.2 and 3.3. Suzanne prepared the datasheets which appear in chapters 2.2 and 2.5. Rolando, as well as Else Magaard, Jens Christian Svenning, and Constanza Rios wrote methodology pamphlets in Spanish which helped me a great deal in writing this longer text. Suzanne Lao also collected information on equipment we used in the BCI census. Stephanie Bohlman, Elizabeth Losos, Jim LaFrankie, Mary Schultz, Jill Thompson, Jens Christian Svenning, and Else Magaard offered many comments on the manuscript as it was being prepared. Finally, I thank Tobias, Nat and Luc for only occasionally pestering me while I worked at home on the book, and Mary Schultz for extensive editing of the manuscript.

Finally, I thank all the funding organizations that have made The Center for Tropical Forest Science possible. The John D. and Catherine T. MacArthur Foundation provided a major, long-term grant that was instrumental in extending the plot network from three to twelve sites. Other important contributors have been and are: the Smithsonian Tropical Research Institute (Panama), the Smithsonian Institution (U.S.), the National Science Foundation (U.S.), the Forest Research Institute of Malaysia, the Sarawak Forest Department (Malaysia), The Andrew W. Mellon

Foundation (U.S.), the Tupper Trust (U.S.), the Indian Institute of Sciences (India), The John Merck Fund (U.S.), the Wildlife Conservation Society (U.S.), the W. Alton Jones Foundation (U.S.), Instituto de Investigaciones de Recursos Biológicos Alexander von Humboldt (Columbia), the Royal Thai Forestry Department (Thailand), the United States Agency for International Development, the National Institutes of Health (U.S.), Conservation International (U.S.), Isabela State University (the Philippines), the United States Forest Service, the Foreign Currency Exchange Program (U.S.), the Rockefeller Foundation (U.S.), the Southern Company (Japan), the Norwegian Overseas Aid Agency, the Japan Ministry of Education (MONBUSHO), the Keidenren Nature Conservation Fund (Japan), the World Wildlife Fund (U.S.), the United Nations Education, Scientific, and Cultural Organization, the Center for International Forestry Research, the Organization of American States, the Turner Foundation (U.S.), Fundación Natura (Panama), the Avina Foundation (Switzerland), the Conservation, Food, and Health Foundation (U.S.), Centre de Formation et de Recherche en Conservation Forestière (Democratic Republic of the Congo), the National Institute of Education and the Nanyang Technological University (Singapore), the Bioresources Department and Conservation Programme (Cameroon), the University of Peradeniya (Sri Lanka), Harvard University, Yale University, Princeton University, Osaka City University (Japan), the University of Kyoto (Japan), Ehime University (Japan), the University of Puerto Rico (U.S.), Pontificia Universidad Católica de Ecuador, Oregon State University (U.S.), the University of Aarhus (Denmark) and the Singapore National Parks Board.

CHAPTER 1.2

Design and Purpose of the Large Plot Network

Developing a long-term and large-scale study of a tropical forest requires careful consideration of the site and the personnel. In order for me to present general recommendations on these considerations, I first present the overall rationale for the CTFS network of large plots and explain how plot locations were chosen. Then I offer advice on where a large plot might be placed and how many people are needed to run it.

The Purpose of a Large Forest Plot

The goal of a large-scale tropical forest plot is to provide demographically useful samples of a large number of co-occurring tree species. Because tropical forests are diverse, with hundreds of tree species sharing a community, there is no way to study the demography of even half of the species except by censusing a very large number of stems. A single large plot is a comprehensive and precise way of sampling a large number of stems.

Large plots are also comprehensive in their inclusion of saplings down to 10 mm in diameter. This provides demographic data on juveniles as well as adult trees. With long-term data on the density and spatial distribution of adults and juveniles of a large number of species, demographic models that incorporate spatial variation and density-dependence can be developed. Each plot site then becomes a focus of basic research on many different topics, including community ecology, genetics, physiology, phenology, and animal-plant interactions.

This basic information has many applied spin-offs. Demographic information provides precise data on the productivity of economically valuable timber or non-timber species, and can thus be used to develop models of sustainable extraction. Similar models can be used to evaluate other economic values of the forest, such as carbon storage or species preservation. Long-term data can be used to monitor changes caused by climatic shifts or human disturbances, and evaluate extinction rates of tropical tree species. In addition, data on individual species can be used to help propagate trees for reforestation or plantation forestry.

But a large-scale plot in a tropical forest is a major investment. A single plot requires a great deal of effort on the part of multiple scientists; however, it yields vast amounts of information applicable to a wide variety of questions. As a scientific project, a large tropical forest plot is akin to a space-station or a probe to Mars—it is a comprehensive scientific tool used jointly by a large number of scientists from several disciplines. And, like other big science projects, it is costly.

Tropical Forest Census Plots: Methods and Results from Barro Colorado Island, Panama and a Comparison with Other Plots, by Richard Condit.

The CTFS Network of Plots

As large and expensive tools for studying tropical forests, these forest-dynamics plots must be placed carefully. The goal of the Center for Tropical Forest Science (CTFS) is to select a few key sites for large plots, in an attempt to include each major forest block and climatic type of the tropics (Condit 1995). In Asia, there are now large plots in dry forest (India), seasonally moist forest (Thailand), and aseasonal, everwet forests in Malaysia (two sites), the Philippines, and Sri Lanka (Table 1.2.1). The sites in Malaysia contrast in soil fertility, while the other two sites represent insular forests, one affected by typhoons (the Philippines). A parallel network is under design in Latin America, where a Puerto Rican site in the Caribbean is similar to the site in the Philippines, the Barro Colorado site in Panama parallels the seasonal Thai forest, and the plot in Ecuador is similar to the plot in Peninsular Malaysia (Pasoh). Sites comparable to Mudumalai in India and Lambir in Sarawak are under discussion. In addition, there is a montane plot in Colombia which has a recent parallel in the mountains of southern Thailand (not listed in Table 1.2.1). A similar program is also underway in Africa (Table 1.2.1).

The plots range in diversity from the dry forest in India, which has 71 species of trees and shrubs in 50 ha, to extraordinarily diverse forests in Ecuador and Malaysia (Table 1.2.1). The Lambir plot has 1175 tree and shrub species in 52 ha, far more than can be found in all of the United States and Canada (Elias 1980).

Most of the existing plots are in undisturbed forest with protected populations of native animals. The Mudumalai and Luquillo plots have been logged, though, and several of the Asian plots have seen significant poaching of vertebrates. CTFS is currently also considering large plots designed specifically to examine human disturbances. One plan is to set up a replicate pair of plots in an area of timber extraction, with one large plot in logged forest and another in fully protected forest. This is a new direction and requires a somewhat different approach, nevertheless, the basic methods would be followed—sampling all trees and shrubs ≥ 10 mm diameter—and the fundamental purpose would be the same—to study the demography of a large number of individual tree species at one site.

In any case, no one envisions using the large plot tool at more than a few key sites. There will probably never be more than 30 in the world. For this reason, new sites for these projects should be chosen carefully, with an eye toward meshing with the larger network. The limited number feasible also demands that existing sites be carefully standardized, and this is the primary purpose of this book.

Selecting a Site

The overriding issue in selecting a site for a new large plot is relevance to an important scientific and management question, which in turn fits into the larger objectives of the CTFS network. Beneath these broad goals are a series of more mundane practical considerations. A hierarchical list of the criteria that should be considered in site selection follows:

Biogeography

First and foremost, the large plot network is designed to cover the major tropical forest regions of the world, and sites should be typical of broad areas in each biogeographic realm. New sites that expand the CTFS network to important regions not yet under study are a priority, for example, new sites in Latin America and Africa, where plot networks are incomplete.

Table 1.2.1. ***The CTFS network of large forest plots. Blanks indicate sites where data are not yet available.***

country	site	species in plot	individuals in plot	area (ha)	shape (m)	rainfall (mm)	dry season (mos.)	current status
Panama	BCI	300	230,000	50	1000x500	2500	4	4 censuses done
Puerto Rico	Luquillo	120	80,000	16	500x320	3500	0	2 censuses done
Ecuador	Yasuní	~1100	~310,000[1]	50	1000x500	2500	0	17 ha done, planned completion in 1999
Colombia	La Planada			25	500x500	3500	0	just underway, planned completion in 1998
Peninsular Malaysia	Pasoh	816	335,000	50	1000x500	2000	0	2 censuses done
Thailand	Huai Khae Khaeng	247	100,000	50	1000x500	2000		1 census done but dataset not complete
Sarawak, Malaysia	Lambir	1175	350,000	52	1040x500	2600	0	2 censuses done but data set not complete
Philippines	Palanan			16	400x400	3500	0	8 ha done, completion pending
India	Mudumalai	71	16,000[2]	50	1000x500	1200	6	3 censuses done[3]
Sri Lanka	Sinharaja	209	~180,000	25	500x500	3500	0	1 census done
Cameroon	Korup	~500[4]	~175,000[4]	25	500x500	5500	3	just underway, planned completion in 1998
Dem. Rep. of the Congo	Ituri	~450[5]	~300,000[5,6]	40	4 plots 200x500	1700	2	1 census done,data set not complete

[1]Species count extrapolated from first 2 ha, individuals on first 11.5 ha; [2]26,000 in first census, but declining; [3]Plus seven partial censuses (annual); [4]Extrapolated from first ha; [5]Not including lianas; [6]Extrapolated from first 10 ha

Local Site Conditions

The second level of criteria applies to climatic, soil, and disturbance features within each biogeographic realm. Sites that are broadly representative of important variation in these characteristics are sought; examples of climatic variation in the existing network were described in Table 1.2.1. Current priorities for CTFS include a dry forest plot in Latin America and a plot at forest-savanna boundary in Africa, and a plot with a short dry season in Asia.

Host-Country Partner Institutions

Having local scientific participation in a large plot is crucial. A major goal for CTFS is to assist local institutions in developing a research tradition, so that they can extend and expand a large plot on their own. In addition, it is usually local scientists who know the forests and the flora, and local collaborators facilitate political aspects of the project (e.g., permits and hiring decisions). Host-country partners also provide the main project office, where plant specimens and data are stored and where data entry and analysis takes place. Identifying capable partner institutions is thus a key initial phase in a project.

Science and Management Objectives

Although data from large, long-term plots will have many scientific and practical uses, key research topics which form the backbone of the project should be specified in advance. Scientists or land-managers from the host country are familiar with local issues for which a large forest-dynamics plot would provide relevant information. An example of a new research direction for CTFS was given above: a comparison of logged and unlogged forests. This project will be developed at a site where the topic is considered a priority for research. Of course, there are many other issues which might provide underlying scientific goals.

Logistics and Feasibility

There are a number of more mundane issues which must also be considered. Since a great deal is invested in setting up a large plot, it is crucial that there be adequate protection. Accessibility is also an issue, since a large plot requires a substantial labor force (described below), and it is difficult to have the entire crew living permanently at the site. Commuting to the site is thus preferable, either daily, weekly, or monthly.

Except where a daily commute from a nearby town to the plot site is possible, there must be living quarters for the work crew near the plot. Local office and laboratory space are also important for preparing daily work, storing data sheets, and storing and sorting plant specimens. Herbarium cabinets are needed. If electricity is available, photocopying and some computer work could also be done at the plot site. Living quarters and a local office may have to be built specifically for the project.

Personnel

A large forest plot in the tropics should be viewed as a big, interdisciplinary effort, and the work force must reflect this. There are five general categories of labor needed in the project: the scientists who manage it, field supervisors who oversee the field work, the main labor force of field technicians, a database manager to organize the computer work, and clerical personnel who transcribe field data.

Scientific Managers

These are the principal investigators who devise the project and set the goals. Ideally, there should be at least three, including one taxonomist to manage the species identification and one quantitative ecologist or forester to manage the data. At least one of these managers must be from the host-country. The scientists must be people who consider the large plot project *their* work, and who will take complete responsibility for its outcome. They must have a vested interest. Successful plots are usually where local scientists take the initiative to start a collaborative and standardized project, rather than where foreign scientists initiate the project by seeking local partners.

The scientists' primary responsibilities include raising funds for the project, choosing the site, devising the research protocol, hiring and training the field supervisors, initiating major equipment purchases (e.g., vehicles, copiers, computers), assuring that permits are in place, finalizing the database after work is completed, and writing up the results. An administrator should be hired to handle some of the logistic details.

Field Coordinators

Field managers should be hired as soon as possible, even before the site is selected. They need to be experienced field technicians, preferably natives to the host country and familiar with the forest and the people. There are two jobs for the coordinators: overseeing the surveying and tree mapping, and carrying out the taxonomic work. Depending on the site and the flora, two coordinators may handle both tasks, or two pairs may be needed, with two people on each task (the latter where taxonomic work is especially difficult due to limited prior knowledge of the flora).

Field Assistants

These are the people who do the bulk of the field labor. They should enjoy working in the forest, but this position does not require prior experience. About 12 assistants should be counted on for finishing a 50 ha plot in three to four years.

Database Coordinator and Clerical Assistants

Two people working at the main office will be necessary for entering data from field sheets into computers. A third person who manages the computer work—data entry and construction of the main database—is also needed, but this probably does not require full-time attention. Thus the database manager could also have other responsibilities, perhaps helping plot administration. The database manager, of course, must be skilled with computer hardware and software.

Further details of how these coordinators and assistants complete a 50 ha tree inventory are provided in this guide. In Part 2, I provide details on the field work for the large census, and in Part 3, details on the database work. The last section of the book, Part 4, illustrates plot results with distribution maps for select species.

Part 2:
Field Methods for a Large Forest Plot

Part 2

Introduction and Overview

Here I present seven chapters on field methodology for a large and long-term forest inventory. Chapter 2.1 describes the surveying methods for gridding the plot, which must be done prior to tree mapping. The heart of Part 2 is chapter 2.2 on the field census, which describes the tagging, mapping, and measuring of trees. Chapter 2.3 is on tree identification, which warrants separate coverage because it is usually done separately from the mapping, owing to the difficulty inherent in tropical tree taxonomy. Chapter 2.4 is on error correction—a key phase of the study since errors are inevitable in such a large dataset: enormous time and effort can be saved by catching them early. Chapter 2.5 is on the recensus of a large plot, and focuses on issues that arise for the first time in a second census. Chapter 2.6 rounds out the methods section with a short description of additional censuses that may be added to the main census, and chapter 2.7 is a summary of the entire field section, including an estimate of the total effort—time, labor, and cost—required.

My strategy in presenting the field methodology is to first present a detailed series of recommendations on how a large forest census in the tropics should be done. These are not necessarily the same as the methods used at BCI: I did not want to open each chapter recounting out-of-date BCI methods which I no longer consider satisfactory. Instead, I present first an idealized methodology which corrects these errors and which takes into account my experience with other scientists working on other plots.

After presenting field methods, I describe the rationale for the recommendations as well as alternate methods followed at different plots. I include in this section exactly what has been done at BCI, even when I now consider it incorrect; I thus attempt to fully document the BCI methods, but explain where necessary why they do not match my current recommendations. Most important, by providing the rationale behind all methods, I hope to indicate where deviation in methodology is reasonable and where it is not—that is, which issues are critical for data standardization, and which are a matter of preference. Forest biologists have their own methods and will probably continue to use them; however, by explaining the rationale behind our standardized methods, I provide the option for other workers to make their own decisions about petty details of data collection while still achieving comparability. Indeed, experienced ecologists and foresters may see ways to improve my methods while maintaining standardization.

Tropical Forest Census Plots: Methods and Results from Barro Colorado Island, Panama and a Comparison with Other Plots, by Richard Condit.

CHAPTER 2.1

Topography and the Mapping Grid

Establishing an accurate and permanent grid system allows trees to be mapped precisely and relocated easily, and is also the basis for developing a topographic map of the plot, providing elevation, slope, and aspect information—all of which are crucial for understanding tree distributions. Indeed, most of the really interesting data emerging from the large plots is based on accurate spatial positioning, which, in turn, is based on the grid system. Surveying in forest is difficult, more so in a vegetation plot because the plants *must* not be damaged. The task of establishing a very accurate mapping grid can be time-consuming, but is crucial for obtaining high-quality data.

Here I present a methodology for surveying the grid and creating a topographic map of a forest plot. This includes laying out a 20-m grid over all 50 ha, with elevation established and permanent markers placed, then adding smaller markers on a 5-m grid. The recommendations complement and extend the report of Manokaran et al (1990) on the Pasoh plot. Following the recommendations, I briefly describe alternative methods that have been followed at other plots, along with a rationale for the recommended methodology and the alternatives.

Equipment and Supplies

Surveying Tools

A surveying compass along with a durable but flexible tape measure will suffice. The compass consists of a large horizontal compass face and needle, on top of which sits a telescopic sighting device which can be rotated horizontally to take a bearing and vertically to measure inclination (Forestry Supplier sells a Sokkia surveying compass and tripod, catalogs #37485 and #37487). Two sighting poles are also necessary: approximately 2 m in height with tape measures attached at identical heights on each. One pole should have a leveling bubble or plumb bob so that it can be held precisely vertically. Rubber straps or ropes are also needed for pulling and tying vegetation out of sighting lines.

Grid Markers at 20-m Corners

The mapping grid is marked permanently. In most circumstances, I recommend use of PVC tubes as markers: 1.2 m long by 50 mm in diameter, with a 1.5 mm wall thickness. It is useful to paint the top orange to enhance visibility. Stakes must be sunken as deeply into the ground as possible, as the goal is to have them survive tree falls and herds of passing mammals. Other options for markers are discussed below under alternative methods.

Tropical Forest Census Plots: Methods and Results from Barro Colorado Island, Panama and a Comparison with Other Plots, by Richard Condit.

Two or three bright pieces of orange flagging, about 1-m long, should then be hung from tree branches around each stake. This makes the 20-m corners easy to find from a distance. The coordinates of the 20x20 m quadrat should be marked on the flagging and on the top of the stake (the coordinate system is described below). These flags need to be replaced every few years. Forestry Suppliers sells various flexible, colored, plastic flagging material.

Grid Markers at 5-m Corners

These should also be PVC stakes, when possible, but smaller than the 20-m markers: 1 m long with a 13 mm diameter and 1.5 mm wall (it is important that they be readily distinguishable from the 20-m stakes in the field). Ideally, they should be just as permanent as the 20-m markers, but this has not been the case at all large plots, and I discuss options for the 5-m markers below.

Phase 1—The Survey

Surveying a Line

To survey a single line means placing a series of stakes at precisely 20-m intervals along a straight line. At least three and usually four people are needed for this. One person (Surveyor A) handles the surveying compass, setting it directly above a grid marker using the compass' plumb bob and aiming it in the chosen compass direction. It is convenient, but not necessary, to make this a cardinal direction. A second person (Surveyor B) runs the tape measure through the vegetation. The third and fourth people move vegetation aside to allow a sighting line from the surveying compass to a point exactly 20 m ahead. Saplings and lianas can be bent aside and tied in position with rubber straps or ropes. Vegetation must not be cut!

Surveyor A (A) then measures the height above the ground of the sighting telescope using one of the sighting poles. Surveyor B (B) holds the other sighting pole vertical, close to the 20-m marker on the tape measure, and moves it from side-to-side until Surveyor A finds the pole through the telescope. Surveyor B then indicates with his or her finger the height on the second pole that matches the height of the telescope as measured by A. Shining a flashlight on the finger can help in dark forest. Surveyor A must locate the finger with the cross hairs in the telescope and note the inclination (vertical angle).

If the inclination is non-zero, a distance correction is necessary, because the 20 m must refer to horizontal distance (Fig. 2.1.1). The distance d, parallel to the ground, is longer than 20 m, and is found from the formula

$$d = \frac{20}{\cos\theta} \qquad \text{(eq. 2.1.1)}$$

where θ is the inclination. For example, if $\theta = 10°$, d is 20.3 m. Surveyor A should carry a table of d-values for θ in half-degree increments from 0.5° to the steepest slope that might be encountered.

With the preliminary reading of θ, a corrected value for d is read from the table. The distance must be measured precisely: the tape measure is pulled taught and held at the same height at both ends; any vegetation deflecting the tape is moved. Surveyor B moves the sighting pole back to precisely d meters, and A must relocate the pole (asking B to move to the side if necessary) and read θ again. If changed, a new d must be read from the table and the pole moved again and resighted. When θ does not change, B places a permanent marker at the spot.

Surveyor A records the horizontal distance (usually 20 m), the inclination θ, the bearing, and the numerical designation of the two stakes (the numbering system is described below). It is important to record the inclination as negative for downward slopes and positive for upward; this is easy to forget! Also, the stake numbers must be recorded in the correct order—first the stake where the surveying compass stood and second, the one toward which it was aimed.

After the first two stakes are placed, the third stake can be placed on the same line easily. The surveying compass is placed on the second stake and aimed back at the first. The compass then allows the telescope to be rotated exactly 180° (or any other angle), without referring to the needle; there is a precise vernier scale for this purpose. The compass is rotated 180°, then the third stake placed on this line using the procedure described above. Stakes to the side can likewise be placed at 90° and 270°. This simplifies the surveying, but by eliminating use of the needle, errors in direction could accumulate. Thus, I recommend that after every five stakes on a line, the needle be used to set the direction. If it has drifted off by more than 0.5°, prior sightings should be re-checked with the needle.

Placing the Central Axes of the Plot

The initial survey lines define the plot's shape and must be precisely placed. In order to ensure their accuracy, they should be shot in sets of three parallel lines, 20 m apart. One set runs the plot's long axis (1000 m for a 50 ha plot) and a second runs perpendicular to this. It is best to make these lines the central axes of the plot, since this reduces the distance to all subsequent grid markers (Fig. 2.1.2).

First, the surveying team should shoot two points, 20 m apart, along one line. In the example illustrated in Figure 2.1.2, this line goes north. Then a parallel line should be initiated 20 m to the west, and a second parallel line 20 m to the east. The three parallel lines are extended 20 m at a time, and sideways sightings are taken between every adjacent pair (Fig. 2.1.2). If side-sightings yield discrepancies > 20 cm or > 0.5°, then the original sightings should be re-done. The inclination measured from side-sightings should be recorded, as it will contribute to the elevation map. Side-checks prevent the three lines from "drifting" away from parallel, as would certainly happen in a 1000-m line.

Unfortunately, in a real forest, large or fallen trees block some sighting paths. When one of the northward lines is blocked, it is necessary to place a new stake from one of the parallel lines. Likewise, side-sightings will be blocked, so it is never possible to do side-checks at every stake. For example, in a 10 ha section of the Congo plot, 44% of all 20-m sightings were impossible. As long as errors can be kept small enough that side-checks are only necessary at 40-60 m intervals, this is not a problem. There will inevitably be some stakes that cannot be sighted from any of the four adjacent stakes. Here, a sighting from a diagonal stake must be used, with the distance adjusted to 28.3 m. Intermediate sightings to the mid-point of a quadrat (14.14 m) might be necessary in this circumstance.

Another difficult circumstance is when the position of a new stake is occupied precisely by a tree. If the trunk has a diameter < 40 cm, I recommend putting the stake as close as possible to its real position; at this diameter, it will be within 20

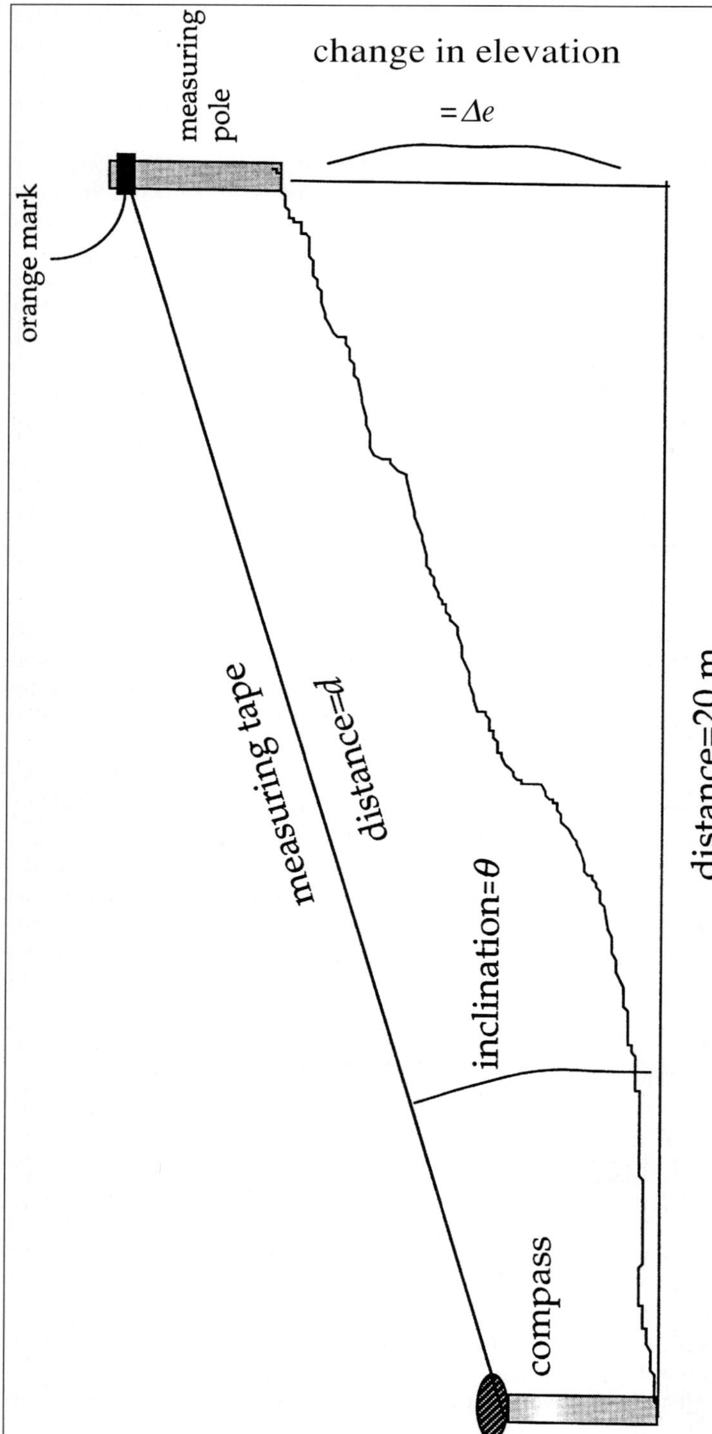

Fig. 2.1.1. An illustration of how the distance between two stakes must be measured. The recorded distance between stakes (usually 20 m) is the horizontal distance. The measured distance, d, is parallel to the ground and always $\geq$ 20 m (= 20 only if there is no slope). The change in elevation is Δe and the angle of inclination is θ. See equations 2.1.1 and 2.1.2 in the text.

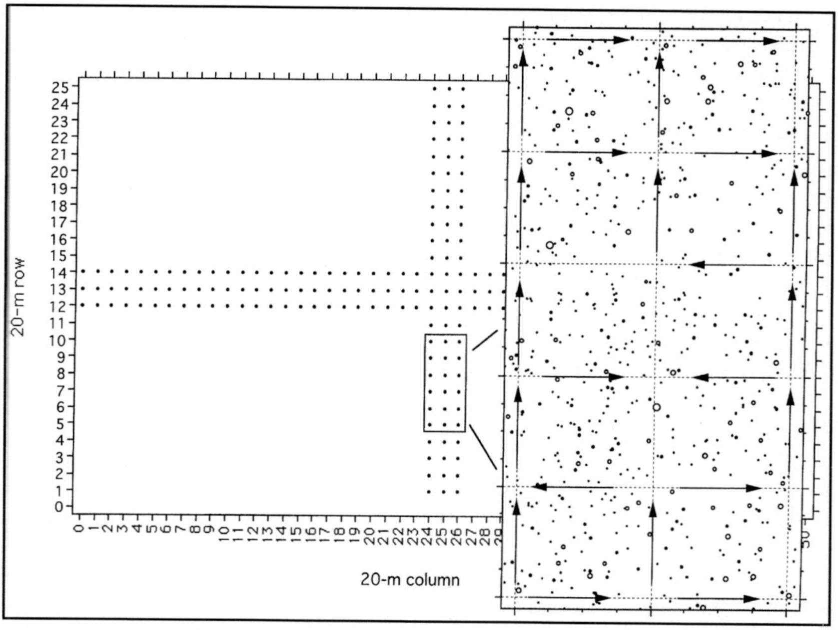

Fig. 2.1.2. Map of a 50 ha plot, showing row and column numbering, and an idealized scheme for surveying the initial lines of the plot. North is upward. The large map shows a rectangular region 1000x500 m. Points are 20 m apart, and 20-m rows and columns are numbered along the axes. The three parallel lines along central axes indicate the first lines that should be surveyed to set the plot. The inset shows in detail how a 100 m section would be surveyed. The circles on the inset represent trees, with locations and sizes taken from a real section of the Barro Colorado plot. All trees ≥ 40 mm in stem diameter are marked. Stems < 40 mm can be bent aside and thus do not block sightings. Black arrows indicate where sightings would be taken. Vertical arrows are the main sightings along the three parallel lines, and horizontal arrows are side-checks; 4 of the 27 sighting paths are blocked by trees. This is a hypothetical surveying scheme, not the real scheme used in this portion of the BCI plot.

cm of the correct position. For larger trees, the PVC stake should simply be attached to the trunk with a nail, and the flagging hung as usual to make it visible. In either case, the stake should not be used as an origin of sightings to other stakes, as this would compound the error. If a large rock occupies a stake's position, paint can be used to mark the location.

Figure 2.1.2 shows a fictional but plausible surveying record for a portion of the initial axes. A real record is shown for a 4 ha section of the Congo plot in Figure 2.1.4.

Surveying the Remainder of the Grid

When the accuracy of the initial axes is assured, the full grid can be laid out following the same protocol. New lines are shot parallel to the existing lines. Frequent side-checks to already surveyed lines are important, but with experience, the surveying team can gauge its own accuracy, estimate how rapidly drift errors accumulate, and reduce the number of side-checks accordingly.

Additional Surveying in Dissected Terrain

If there is a substantial mound, hole, or stream within a quadrat that would be missed by placing stakes only every 20 m, additional points should be surveyed at either 10- or 5-m intervals in order to capture the feature. As a rough guideline, I suggest that features < 1 m in height (or depth) can be ignored. The surveying procedure described above for 20-m points is followed exactly for 5- or 10-m points, with the exception that a table of distance corrections for 5 and 10 m lengths is needed. Permanent stakes should be placed at these positions, but marked carefully to indicate their precise location (to prevent confusing them with the main 20-m stakes).

It is not necessary to survey every 5-m corner within a dissected 20x20 m quadrat, just the minimum necessary to capture the topography. The remainder will be filled in with a tape measure, as described below.

Phase 2—Placing the 5-m Grid

In smooth 20x20 m quadrats, having no topographic features, the four corner posts completely describe the terrain, and 5-m points can be located with a tape measure alone. The tape is first held between two 20-m points, and three intermediate markers are placed to divide the distance into four equal sections (since the distance along the ground is > 20 m on a slope, the intervals may be > 5 m). In this way, three markers are placed between every pair of 20-m points. Subsequently, the nine interior markers are placed by laying the tape between appropriate pairs of the new markers.

In dissected quadrats where additional points were surveyed, some 5-m stakes were already placed by the surveying team. The remaining 5-m posts can be placed with tape measures alone.

Quadrat Numbering

The 20x20 m quadrats into which the large plot is surveyed become the units of field labor, and a simple and consistent numbering system for them is necessary. The most convenient system for field work is to designate a quadrat by column and row, where rows are 20-m swaths numbered from south to north, and columns from west to east. Quadrat 3513 is column 35, row 13 (Figs. 2.1.2, 2.1.3). By arbitrary standard, the initial row and column are 00, thus quadrat 0000 is in the southwest corner, and 0013 is in the western-most column, row 13. In a 1000x500 m plot where the long axis is east-west (Fig. 2.1.3), the northwest corner is 0024, the southeast corner 4900, and the northeast corner 4924.

The stake at the southwest corner of a quadrat has the same number as the quadrat. Thus, the far southwestern stake of the plot is 0000. The southwestern corner of quadrat 3513 is stake 3513. The northwest corner of the same quadrat is stake 3514, the southeast corner 3613, and the northeast 3614 (Fig. 2.1.3). These numbers are marked on the stakes and on flags above them, to allow navigating in the plot. To find any quadrat or a tree within, proceed first to the stake with the same number, then look into the quadrat to the northeast.

If a plot is not placed following cardinal directions, then one direction must be chosen as the major plot bearing, replacing north in Figures 2.1.2 and 2.1.3. Quadrat numbers are assigned as described above, but with the map rotated so this main bearing is up (i.e., quadrat 0000 is in the bottom-left, etc.). Stake number

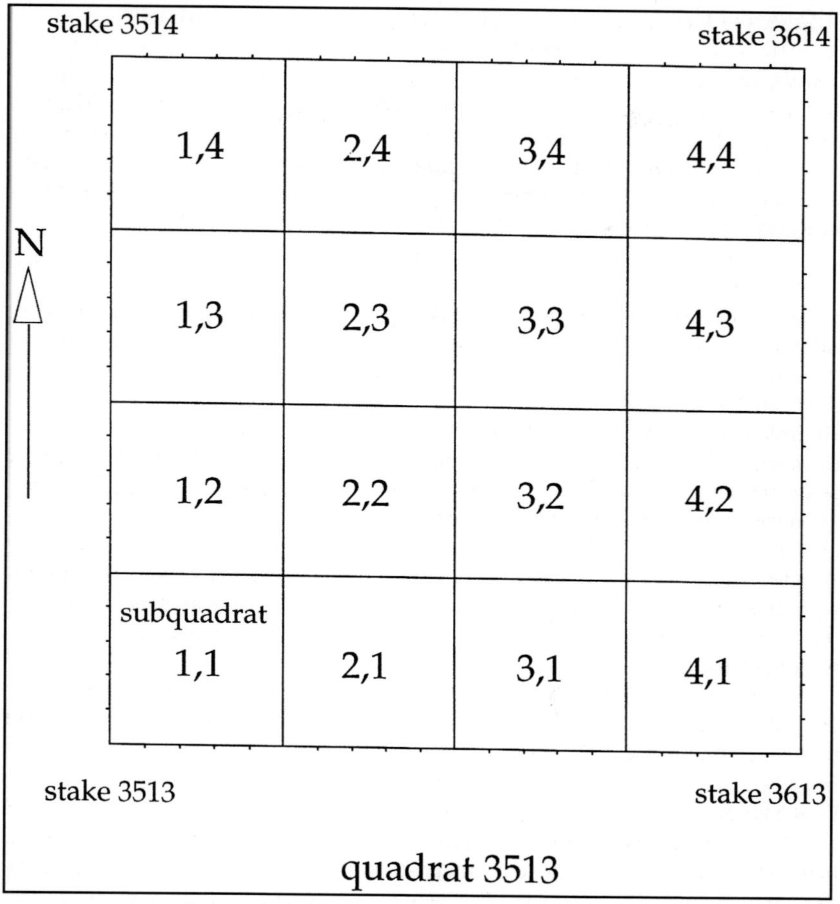

Fig. 2.1.3. Numbering of 20-m stakes and quadrats and 5x5 m subquadrats.

3513 lies at the bottom-left of quadrat 3513, by convention. Navigating in the plot involves standing at a given stake and facing the direction of the main bearing; then row numbers increase ahead and column numbers increase to the right.

The 16 5 x 5-m subquadrats within each quadrat are numbered from 1,1 to 4,4. The first number indicates the column (1-4), and the second number represents the row (Fig. 2.1.3). This is not consistent with the numbering of columns and rows, which starts with 0, but this arbitrary system was set in place at Barro Colorado in 1980 and has been used since that time.

Phase 3—Calculating the Topography and Its Accuracy

The Data

Phases 1 and 2 provide a data set that includes the location of pairs of points and the vertical inclination between many of the pairs. The first step taken with this data is to convert the inclination to a change in elevation: if d is the horizontal distance between stakes (usually, $d = 20$) and θ the vertical inclination, the change in elevation is

$$\Delta e = d \tan \theta \qquad \text{(eq. 2.1.2)}$$

(recall that Δe and θ are negative for downward slopes). Figure 2.1.1 illustrates this relation.

Converting these data to a topographic data set is rather complex, and is best illustrated with a real example. Figure 2.1.4 shows a 4 ha sample of elevation measurements made at the Ituri plot in the Congo. Points are every 20 m, and the arrows between points indicate the direction of survey sightings. The numbers next to the arrows indicate Δe along each sighting, in centimeters. Rows and columns are numbered as described above, starting at 00 in the lower left. Table 2.1.1 shows a portion of the raw data set; in the table, points are numbered as a sequence of integers, rather than with row and column designations. (The table's legend explains the change in numbering—the change is useful for the analysis.)

Calculating Topography

The goal is to come up with an estimate of the elevation of each point, given estimates of the change in elevation between adjacent points. One corner of the plot is arbitrarily set to elevation 0, and the elevation of each point relative to this must be calculated.

The difficulty is that side-sightings mean that many points will have several sightings to them, evident in Figure 2.1.4 as points where multiple arrows meet. When this is the case, there will be alternate estimates for the elevation of a point. For example, if stake 0000 in the figure is set at elevation 0, stake 0001 is –79 cm, 0002 is –85 cm, and 0101 is –12 cm. Then there are two alternative estimates for 0102: from 0101, it would be +19 cm, but from 0002, it would be –18 cm. The discrepancy of 37 cm reflects error in estimating the inclination angles, and thus error in the elevation map.

Given several alternative estimates for elevation at each point, how can a single best estimate for the entire topography be found? Here I provide a technique for arriving at the optimum (least squares) solution. Consider a point p which has n neighboring points to which (or from which) sightings were taken. Designate the elevation of point p as e_p, and the elevation of each neighbor i as e_i. Then let the change in elevation between point i and point p be Δe_{pi}, being sure to keep the signs straight, so that Δe_{ip} is positive if p is higher than i, and $\Delta e_{ip} = -\Delta e_{pi}$. Then one estimate for the elevation at point p is simply

$$e_p = e_i + \Delta e_{ip} \qquad \text{(eq. 2.1.3),}$$

with one such estimate for each neighbor i (only if a sighting was taken from that neighbor). To get an estimate for e_p based on all n neighbors, just take the average:

$$e_p = \frac{1}{n} \sum_{i=1}^{n} \left(e_i + \Delta e_{ip} \right) \qquad \text{(eq. 2.1.4).}$$

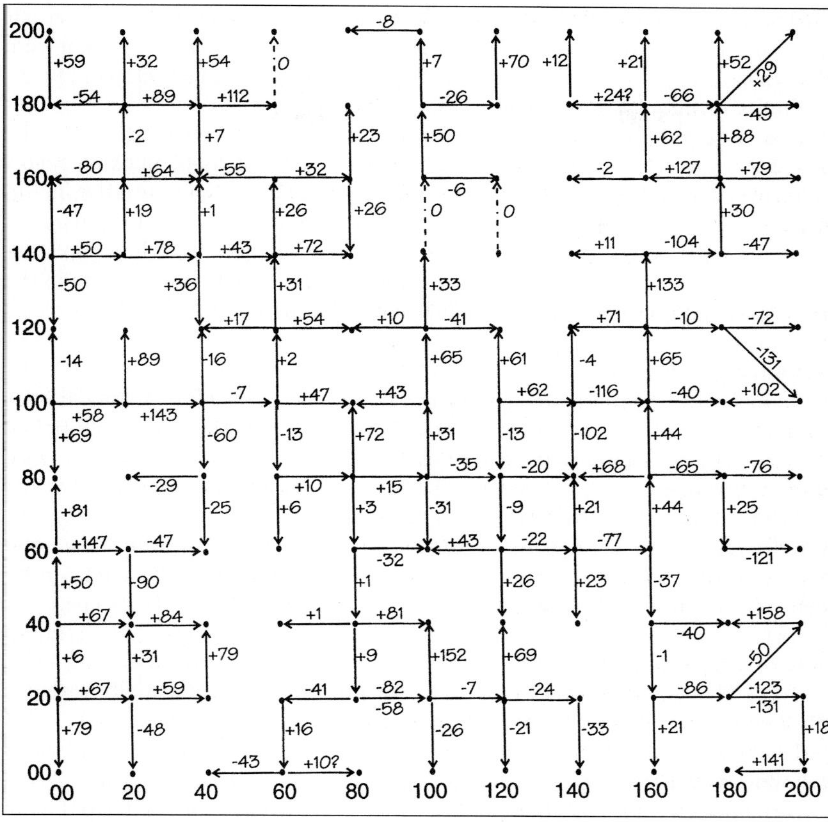

Fig. 2.1.4. A map of 4 ha of the plot in Ituri, the Congo, indicating how surveying was done. Arrows indicate where sightings were taken, with the elevation change next to the arrow. (Data for three of the stakes were missing, and dashed arrows are placed at those stakes; zero elevation change was assumed in those places.)

It is easy to show that this is the least squares estimate for e_p in that it minimizes the sum of the squares of the differences between observed and estimated Δe_{ip}s.

Writing down an equation 2.1.4 for each point p yields a series of N simultaneous, linear equations for the N points in the entire grid. There are N unknowns, one for the elevation of each point. These equations are not independent though, and in fact, the sum of any N–1 of the equations will yield exactly the last equation. To solve the set, one equation must be removed, and the corresponding unknown must be set to an arbitrary value. In practice, I remove the equation for stake 0000 and set the elevation there to 0. The solution to the remaining equations is the set of elevations e_p relative to the one point which was arbitrarily set to 0.

Notice that this approach estimates elevations without any information about the location of the points. It can thus be used for all the 20-m grid points that were surveyed, plus any 5- or 10-m points within. Drawing the maps (below) requires the points' locations, but not the elevations.

In a large plot, solving equation 2.1.4 for each grid point is computationally intensive. This can be done by diagonalizing the matrix of coefficients for the e_i, but fortunately most of the entries are 0, so a program for doing so is not difficult. I have written one such program in C++. The program must read the data set listing the grid points and the elevation change between them, exactly as shown in Table 2.1.1. The output of the program is the elevation estimate at each point in the grid; for the sample data from Ituri, the results are shown in Table 2.1.2.

Estimating Error

Because side-sightings provide alternate elevation estimates for certain points—as explained above for stake 0102 (see *Calculating Topography*)—the accuracy of the estimated topography can be approximated. A simple way to do so is to trace alternative routes between points along sighting pathways. Along each pathway, a total elevation change can be summed, and alternate routes to the same point provide alternate estimates. In the example given above, two routes from stake 0000 to stake 0102 yielded a discrepancy of 37 cm in elevation. Consider a longer example from Figure 2.1.4. From point 0005 to 0308, two completely independent routes can be traced: one straight north to 0008 then east to 0308, the other straight east to 0305, then straight north. The first yields a rise of 188 cm, the second a rise of 253 cm. This can be repeated for longer and longer routes, and for multiple routes of the same length that share no sightings in common.

To illustrate how this provides an estimate of error for the whole plot, I made the following calculations in one 10 ha plot at Ituri. For points at opposite sides of a 20x20 m square, the mean discrepancy was 16 cm (sample of 10); for points at opposite sides of a 40 x 40 m square, it was 28 cm (sample of 10); for a 100x100 m square, it was 96 cm (sample of 4); and for the entire plot (500 x 200), it was 194 cm (only one sample possible). Thus, the estimated accuracy of the elevation change was ± 1.9 m; the total estimated elevation change was 14 m.

Absolute Elevation

Relative elevations of all points in the plot can easily be converted to elevation above sea level if the absolute elevation is available (e.g., from a published topographic map). The absolute elevation at one corner of the plot is simply added to the relative elevation of every point in the plot. At Ituri, for example, the best estimate of the plot's elevation is 750 m, so I would add 750 to all the numbers in Table 2.1.2.

Drawing a Topographic Map

A variety of software packages will draw a map of contour lines based on input such as that in Table 2.1.2. A program called Spyglass for the Macintosh drew the map in Figure 2.1.5A. Topo maps of 1 to 50 ha drawn with elevation at 20 m intervals usually have jagged contour lines, as in Figure 2.1.5A. Most packages can draw smoother curves by filling in elevation estimates using interpolation. The contour

Table 2.1.1. A portion of the sample dataset of elevation changes, corresponding with those shown in Figure 2.1.4. Here the grid points (stakes) are re-numbered as a sequence of integers: stake 0000 in the figure is point 0 here, stake 0001 is point 1, and so on up the leftmost column so that stake 0010 is point 10 here; points 11 through 21 here are column 01 in the figure; etc., with points 110-120 here matching column 10 in the figure. Numbering the points consecutively makes it easier for a program to work with the data to calculate elevation. The entire dataset has 152 lines, equal to the number of sightings made in the 4 ha region (the number of arrows in Fig. 2.1.4); just 17 of those lines are shown here.

Stake 1	Stake 2	Change in elev. (meters)
2	3	0.50
2	1	0.06
2	13	0.67
1	12	0.67
1	0	0.79
5	6	-0.14
5	16	0.58
5	4	0.69
16	17	0.89
16	27	1.43
27	26	-0.60
27	38	-0.07
27	28	-0.16
26	15	-0.29
26	25	-0.25
38	37	-0.13
38	49	0.47
etc.	etc.	etc.

Lines in Figure 2.1.5B were drawn after calculating elevation on a 5-m grid using an interpolation technique called kriging which Spyglass performs easily. If surveying to a 5-m scale is done in the field, no interpolation would be needed.

Rationale for the Methods and Alternative Approaches

Personnel

At most large plots, including BCI, crews have been hired and trained specifically for the plot survey. They form part of the overall field team, and later participate in the tree mapping. But at Luquillo, Yasuní, Sinharaja, and La Planada, trained surveyors were employed to lay the 20x20 m grid. At Yasuní, a team that had been

Table 2.1.2. The least squares best estimate of elevation in a 4 ha region of the Congo plot, based on elevation changes shown in Figure 2.1.4 and Table 2.1.1. The row matches the horizontal axis in the figure, and the column matches the vertical axis.

	column										
row	00	20	40	60	80	100	120	140	160	180	200
200	0.36	0.63	1.66	2.24	2.67	2.75	3.12	3.59	3.44	3.01	2.78
180	-0.23	0.31	1.12	2.24	2.24	2.68	2.42	3.47	3.23	2.49	2.00
160	-0.49	0.40	1.12	1.71	2.01	2.18	2.12	2.68	2.70	1.52	2.31
140	-0.10	0.30	1.07	1.51	2.25	2.18	1.44	2.37	2.26	1.22	0.75
120	-0.60	0.96	1.38	1.25	1.87	1.85	1.44	1.55	0.93	0.87	0.15
100	-0.45	0.07	1.44	1.24	1.70	1.28	0.83	1.50	0.32	0.68	-0.39
80	0.30	0.52	0.81	0.99	0.97	1.05	0.68	0.44	-0.04	-0.69	-1.45
60	-0.44	0.99	0.54	1.05	0.93	0.78	0.57	-0.04	-0.20	-0.44	-1.65
40	-0.83	0.02	0.94	0.70	0.69	1.39	0.68	0.19	0.31	0.39	-0.72
20	-0.83	-0.27	0.24	0.23	0.64	-0.25	-0.16	-0.40	0.71	0.26	-1.14
00	0.00	-0.75	-0.04	0.39	0.49	0.01	-0.37	-0.73	0.92	0.45	-0.96

surveying for oil companies (Yasuní is an area of intensive oil exploitation) was contracted. At Sinharaja, a faculty member in the engineering department at Peradeniya University used the 25 ha plot as a field assignment for a class. In these cases, the surveyors just turned over elevation estimates without any indication of how they were calculated or the degree of error. If this is the case, I recommend re-measuring distance and inclination between selected pairs of points in order to generate estimates of the degree of accuracy of the work.

If professionals are contracted, it is also crucial to emphasize the importance of not damaging vegetation. Experienced surveyors will recommend that accurate work in the forest requires clearing vegetation in order to create long sighting lines. It is possible to survey without doing so, but old habits die hard, so repeating the point may be worthwhile.

Grid Size

All the large forest plots use the same grid dimensions of 20 m and 5 m, and there are good reasons to follow this standard in future plots. For one, both numbers are factors of 100 and are thus useful for most routinely used plot dimensions. For surveying, the 20-m distance between points is close to the maximum distance that can routinely be sighted in forest undergrowth; in the Congo plot, with the densest understory of any of the large plots, sightings of 20 m were possible for 56% of the stakes.

Of course, longer distances could be surveyed in more open forest, and thus it may be more efficient to use a 25 x 25 m grid in some places. There are, however, benefits to using a standardized quadrat size that have to do with data storage and analysis. As will be described in the next chapter and in Part 3, the 20 x 20 m quad-

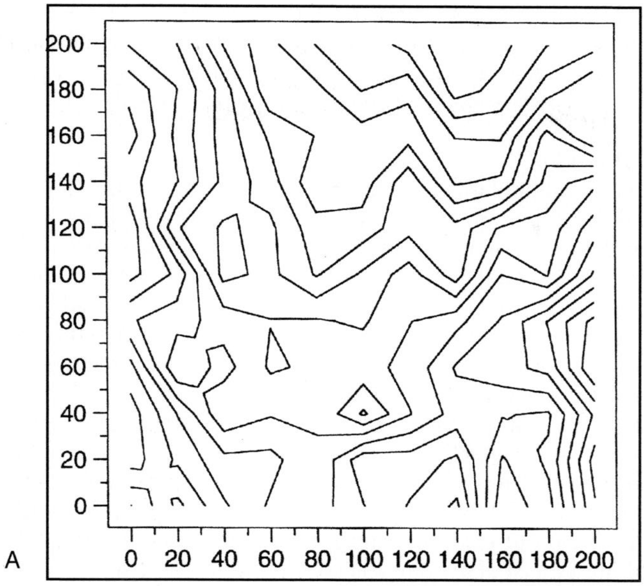

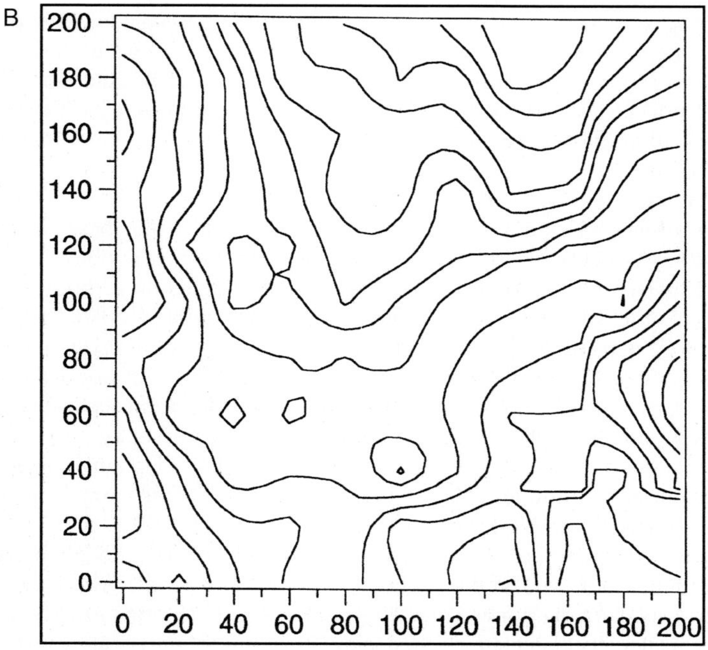

Fig. 2.1.5. Topographic map of a 4 ha section of the Ituri plot, based on the surveying information shown in Figure 2.1.4 and the estimated elevation estimates given in Table 2.1.2. (A) is based on elevation estimates every 20 m. (B) after kriging was used to fill in estimates every 5 m, using Spyglass on the Macintosh.

rat becomes the unit of data collection and entry, and any plot that uses the same quadrat size can use identical computer methods for maintaining the data. This is a matter of convenience, though, not necessity—a plot based on a 25-m grid would still provide data entirely comparable to the existing plots based on a 20-m grid.

Surveying Equipment

The surveying compass was used for marking the plots in the Congo and in Cameroon, and the results appear accurate enough for our purposes. Side-checks suggested that the measurements were repeatable. The obvious alternative is to use a theodolite, which would provide considerably more accuracy (a few seconds of a degree, compared to half a degree for a compass). The main disadvantage to a theodolite is that it is heavier than a surveying compass, and quite a bit more difficult to transport and level in the forest. Theodolites also cannot reference to magnetic north, so directional errors can accumulate.

Tape measures are still the standard for measuring distances in the forest plots, however, there are now relatively inexpensive and extremely accurate laser rangefinders on the market. These would greatly facilitate a forest survey, although I have not yet tried one. Much more expensive are the combined laser-compass surveying stations, and if one of these is available, I would certainly recommend its use in establishing a forest plot—one was used at the Lambir plot (Yamakura et al 1995). They are still very expensive, though. Other advanced technologies for precise mapping, such as satellite systems (Global Positioning System, or GPS) or radio signal mapping, will eventually revolutionize forest surveying, but their accuracy under a tree canopy is not yet satisfactory. Decker and Bolstad (1996), for example, report an accuracy of 4-10 m for GPS under a forest canopy, which is not satisfactory for establishing a 5-m grid.

Grid Markers

What is used to mark the grid corners varies from plot to plot, based on local needs. The key is that the 20-m markers be really permanent. At the BCI plot, aluminum stakes were sunken deeply in the ground and painted orange, but I recommend PVC at other sites because of the fear that aluminum will be stolen (BCI is an unusually well-patrolled tropical forest). Steel is also likely to attract thieves. At Ituri in the Congo, neither PVC nor aluminum was available and wooden stakes were used at the outset. However, the wood quickly rotted, and steel construction bars are being used as replacements. Wooden stakes are not an option.

Stone markers are necessary at Mudumalai in India, where elephants frequently pull stakes out of the ground. This can be a problem with other animals, such as chimpanzees, but so far, Mudumalai is the only plot where stakes could not be used. I would not recommend use of stone markers in general, though, because they are difficult to move and place. On the other hand, if readily available, they can lead to a substantial savings, since PVC or steel stakes are fairly expensive.

Orange flagging has been maintained above every 20x20 m stake at Barro Colorado, with the stake number written on the flagging. Most other plots simply mark the top of the stake itself. When painted orange, stakes are usually visible enough to allow easy navigation (with a compass!) in the forest. The flagging is just an added convenience.

There is more variation among plots in how the 5-m corners are marked. Because of the very large number of 5-m markers needed (18,975 in a 1000x500 m plot, with an additional 1,326 20-m markers), cost, risk of theft, and labor need to be weighed against durability. Both PVC and aluminum are costly: stakes for the entire plot in Ecuador cost over $15,000. And stones, although inexpensive, would require much labor to place every 5 m.

The range of practices for 5-m markers at the large plots follows. At Yasuní in Ecuador, PVC is used. At BCI and Pasoh, 15 cm yellow plastic tent stakes are sunken deeply (Manokaran et al 1990), and 10 mm diameter aluminum canes are placed in contact with them (the stakes for permanence, the canes for visibility). At Ituri in the Congo, there are no permanent markers, instead, tape measures are laid down every 5-m on the day that trees are censused. At Korup in Cameroon, wooden stakes are placed the day of tree mapping, and removed when the quadrat is finished. The drawback to this is that during a second census, if the temporary markers are placed slightly differently, then the location of the mapped trees will appear to shift.

Indeed, this latter point underlines the key to a permanent grid system. Once in place, all future tree mapping is consistent because it is always based on the same reference system.

Surveying Techniques

The key to surveying a large plot in the forest is the use of frequent side-checks as new lines are placed. The rationale for this approach is based on the need to avoid the one error that can crop up in this circumstance: the error of "dead-reckoning". I borrow this phrase from pre-20th century seafarers, who understood well that traveling by compass bearing alone—with no external estimate of latitude and longitude—could lead to huge errors in location (stars provided mariners external latitude estimates, but skies were not always visible). This is simply because a slight deviation in direction over a long distance leads to a substantial deviation in location.

Consider two parallel surveying lines in a 50 ha plot, both aimed due east and starting 20 m apart. A 0.5° discrepancy between the two lines translates into a 9 m error after 1000 m—the lines could be anywhere from 11-29 m apart. Intuition alone suggests that even the best surveyors could not keep the two lines exactly 20 m apart without frequent side-checks. The use of three parallel lines that I recommended above is designed to provide ample side-checks: too many checks cannot be a bad thing.

As long as a compass is used every few stakes to keep the direction of the main line headed east to ± 0.5°, and frequent side-checks keep distance measurements accurate to ± 1%, I contend that the survey is adequate for forestry purposes. Errors will not accumulate, and although the lines may not be exactly straight, area and distance estimates within the plot should remain accurate to approximately ± 2%.

Back-sightings are sometimes recommended to help ensure accuracy. After placing a new stake, the compass is placed on the second stake and aimed back at the first to confirm the direction. I have found these problematic and unnecessarily time-consuming. If side-checks can be done frequently, back-checks are not necessary.

Another important point is that the initial lines be central axes and not the boundaries of the plot. Trying to complete the plot perimeter first would be prone to error, since dead-reckoning would cause the final edge to miss the starting point by several meters. An error of several meters over 50 ha is tolerable as long as it is divided equally across the quadrats, which is what would happen if central axes were placed first and remaining stakes placed outward from them. It is obviously unacceptable to have the entire error appear in the last row of quadrats.

Given this rationale, there are a some minor deviations in the surveying methods described here that are perfectly acceptable. For example, in the Congo, no long axes were ever placed to define the plot. Instead, work started in one corner and proceeded in all directions away from there. Another simple change would be to use distances shorter than 20 m, which might be useful if frequent obstacles prevented longer sightings. Indeed, since stakes will eventually be placed every 5 m, the initial survey could routinely use 5, 10, or 15 m sighting distances depending on the terrain. To date, though, all large-plot surveys have started by placing the 20-m stakes.

Another slight change in method is to measure the vertical change between adjacent stakes, not the inclination. For this, a meter stick is held vertically at the lower stake, and the surveying compass is aimed at the stick and set at zero inclination. This is easy in fairly level terrain, and was the method used at the Ituri plot; it could also be used in steeper terrain with a much longer vertical pole. Another advantage to this technique is that it can allow a crude survey with no more than a hand-held compass and a tape measure, but this would not suffice in areas much larger than a hectare. A disadvantage is the measurements cannot be made uphill, so the compass must always be moved to the lower stake, and this may mean added sightings. In general, unless there is some reason to believe that the protractor scale on the surveying compass is inaccurate, I see no reason not to use inclination instead of height difference.

Surveying the 5-m Grid

In fairly smooth terrain, accurate surveying of the 5-m grid can be omitted. At Pasoh, Mudumalai, Yasuní, and Ituri, no surveying at intervals < 20 m was done. At BCI, the finer survey was only done in 2 out of 50 ha (the far southwest and northeast hectares), regions where small streams are 1-2 m deep and 5-10 m wide. But at the Lambir and Sinharaja plots, which include far steeper terrain, extensive 5-m surveying was done (Yamakura et al 1995).

Time and Labor for Placing the Grid

Based on estimates from several plots, teams of 3-4 people can place 7-10 stakes per day (Manokaran et al 1990). If stakes are placed just every 20 m, this leads to an estimate of 500-700 person-days in the field to finish 50 ha, or, assuming 21 field days per month, 24-33 person-months. The professional crew at Yasuní required a comparable time: it included five people and took five months. These estimates would be revised upward depending on the number of 5-m stakes surveyed. I have no estimate of the time needed to fill in the 5-m grid (without surveying), because this is usually done during tree-mapping.

Key Issues for the Topographic Census

- Lay out first two perpendicular sets of lines on plot axes with extreme care
- Lay out three lines simultaneously so that side-checks can be made between parallel lines
- Keep records from side-checks since they can be used to assess accuracy of the topography.

CHAPTER 2.2

Censusing Trees

The first census of all trees and shrubs is the bulk of the labor in establishing a large plot, but there is a silver-lining: the large effort of marking, mapping, and identifying a quarter-of-a-million plants need never be repeated. Even 20 years later, more than half of those plants will be there, in the same spot, already tagged and identified. Future censuses only add newly recruited saplings, and these will be a small part of the total forest. Thus, the huge chore produces a huge resource that is available for years to come.

In this section I offer a recommended methodology for tree-mapping. This includes locating each stem, tagging it, and measuring its diameter at breast height (dbh). With a grid already in place, locating and tagging trees is straightforward: small squares of land are mapped one at a time. Measuring diameters is more problematic, though, as stem irregularities must be dealt with, and I cover these in detail. Following these recommendations, I provide a rationale for the critical points and a discussion of alternate methods used at various plots. I leave for a later chapter (chapter 2.3) the business of identifying each plant to species, since this requires a different set of skills and a different set of people. It is often done separately from the main mapping.

Personnel

Two field coordinators should supervise the tree-mapping. They must be eager and bright ecologists, foresters, or botanists, with energy and a deep desire to be accurate. Extensive experience is not necessary, indeed, young and enthusiastic biologists (perhaps students) are often the best choice. The two coordinators may be appointed before the field work starts, or they can be identified after the early stages of field work as the best of the mapping crew. They are trained by the chief scientists running the project.

The mapping teams—the mappers—are trained and supervised by the coordinators. Ideally, the mappers will also have some training in ecology or botany, with experience in the forest. However, anyone literate and willing to work in the field can do the job. Indeed, the desire to do precise work is probably more important than experience in ecology or botany (the tree mappers will not be responsible for identifying species; see chapter 2.3). One or two weeks of training should be planned for new workers, measuring and mapping trees under the scrutiny of the supervisors and chief scientists.

Tropical Forest Census Plots: Methods and Results from Barro Colorado Island, Panama and a Comparison with Other Plots, by Richard Condit.

During the first census in most of the large plots, mappers have worked in teams, and although I have argued in the past that working alone is preferable, I now concede that the bulk of the evidence suggests that teams are more efficient. I recommend teams of five, with one person carrying tags to tie to each tree, two carrying calipers and measuring stems, one filling in the map, and the supervisor recording the main data.

Equipment and Supplies

Each mapping team needs three sets of calipers, three diameter tapes (for two measurers plus the supervisor), two 25-m tape measures, a couple small compasses, a hammer, pencils, and two clipboards (for the two people recording data). We get plastic dial calipers and cloth diameter tape from Forestry Supplier (catalog numbers 59591 and 59571); other items are routinely available. Waterproof paper is useful for work in the rain forest, but not essential. If waterproof paper is not used, plastic sheets are necessary to protect the paper from rain. Each measurer also needs a straight pole exactly 1.3 m long, with flat ends, easily made from small trees near (not in!) the plot.

The mapping teams must also place the grid markers at 5-m intervals, and thus need 21 stakes for each 20x20 m quadrat, along with several tape measures for positioning them (as described in the previous chapter). One 40-m piece of light rope is necessary for demarcating each subquadrat.

Aluminum tags are used for marking trees: 44x16x1.3 mm with rounded corners and numbers already engraved (National Band and Tag Co., Newport, Kentucky, USA, tag style #161). Some blank tags should also be purchased to replace lost numbers, and thus a tag punching machine is needed. These are available from Seton Name Plate Co., New Haven, Connecticut, USA (style #M1754). I recommend tying tags to most trees with tough nylon or polyethylene thread. We have purchased 2- or 2.5-mm gauge thread, then unfurled the three braided strands and used each separately. I review a variety of options for tying tags in the section on alternate methods. Aluminum nails, 100 mm long, are needed for attaching tags to larger trees.

Waterproof paint that will last for five years in the forest is needed for marking trunks. Orange Zynolyte fluorescent paint (Zynolyte Products Co., Carson, California, USA) seems to work well, but many oil-based exterior house paints or highway paints probably suffice (Sheil 1995). Each field team needs a paint can and brush.

An aluminum ladder extensible to 5 m is necessary for measuring large, buttressed trees, and large wooden calipers may often be useful (a design is shown in Fig. 2.2.4). A pair of binoculars is also needed when the measuring tape is held high up a large trunk. Since just one team will measure big trees at any one time, a single ladder and one pair of binoculars will suffice.

Plants to Include

At a minimum, the rule for all large plots is to include free-standing woody stems with a diameter at breast height, or dbh, $\geq$ 10 mm. Breast height is 1.3 m above the ground. Rules for locating breast height are given below. Tree ferns and palms are included when their stem is large enough. Many palms have wide enough stems that are not tall enough, so the deciding factor is the location of the top of the stem, and the rule is given below.

In most plots, lianas are not included, but many small lianas appear to be free-standing before they find a host. These should not be included in the census, but the mapping crew will often tag and measure them, not recognizing them as lianas. The tree-identification team should remove these tags (see chapter 2.3). There are occasional species that sometimes grow as lianas, sometimes free-standing; I recommend they not be included (Manokaran et al 1990).

Stranglers also fall in a gray area, sometimes being free-standing (when the host dies), sometimes not. I suggest that stranglers only be included in a census when the host is gone. There are a variety of other marginal species with pseudo-woody stems—bamboos, Marantaceae, Araceae, Cyclanthaceae—that should, in general, be excluded from a large tree census. However, exact decisions about which plants to include will depend on the interests of the local scientists. The rules for handling stranglers, bamboos, and other unusual growth forms have varied from plot to plot. I review alternatives below.

Work Sequence

Each mapping team is assigned one 20x500 m column at a time, starting with column 00. Thus, if there are three teams, they start in columns 00, 01, and 02, and the first to finish one column begins in 03. Work begins at the base of each column and follows the quadrat numbering (Fig. 2.1.2). In a plot laid on a north-south axis, work proceeds northward up each column, and eastward from column to column.

Within each quadrat, teams should follow the same sequence in working through the subquadrats (Fig. 2.2.2). Following a consistent path is useful because the tag sequence is then predictable, and this makes later attempts to find a particular tag easier. It also helps to assure that every stem is located.

Phase 1—Demarcating Subquadrats

Mapping teams are often responsible for filling in the 5-m grid markers, whether these are permanent or temporary. This is best done at the start of work in each 20x20 m quadrat. Placement of stakes was described in the previous chapter.

After the 5 x 5 m stakes are placed, one subquadrat is demarcated with rope. A single piece is wrapped around the four corners to mark the periphery, then across both diagonals to make a large 'X'. The diagonals match those printed on the field map (Fig. 2.2.1A). The rope greatly facilitates mapping, but it must be laid down on precisely straight lines between stakes: if it is off, trees will be mismapped. The rope is moved after each subquadrat is finished.

Phase 2—Data Collection

Data Sheets

Data from each 20x20 m quadrat go onto four types of field sheets: a map, the main data sheet, a multiple-stem sheet, and a problem sheet. Each includes spaces at the top for recording the quadrat number, the mappers' names, and the date on which work in the quadrat begins. Sample data sheets are shown in Figure 2.2.1. One quadrat requires four blank maps, each with a 10x10 m region (Fig. 2.2.1A), and four or more main data sheets (Fig. 2.2.1B). One multiple (Fig. 2.2.1C) and one problem sheet (Fig. 2.2.1D) nearly always suffice for a single quadrat. When data from one quadrat do not fill a sheet, data from a second quadrat should never be

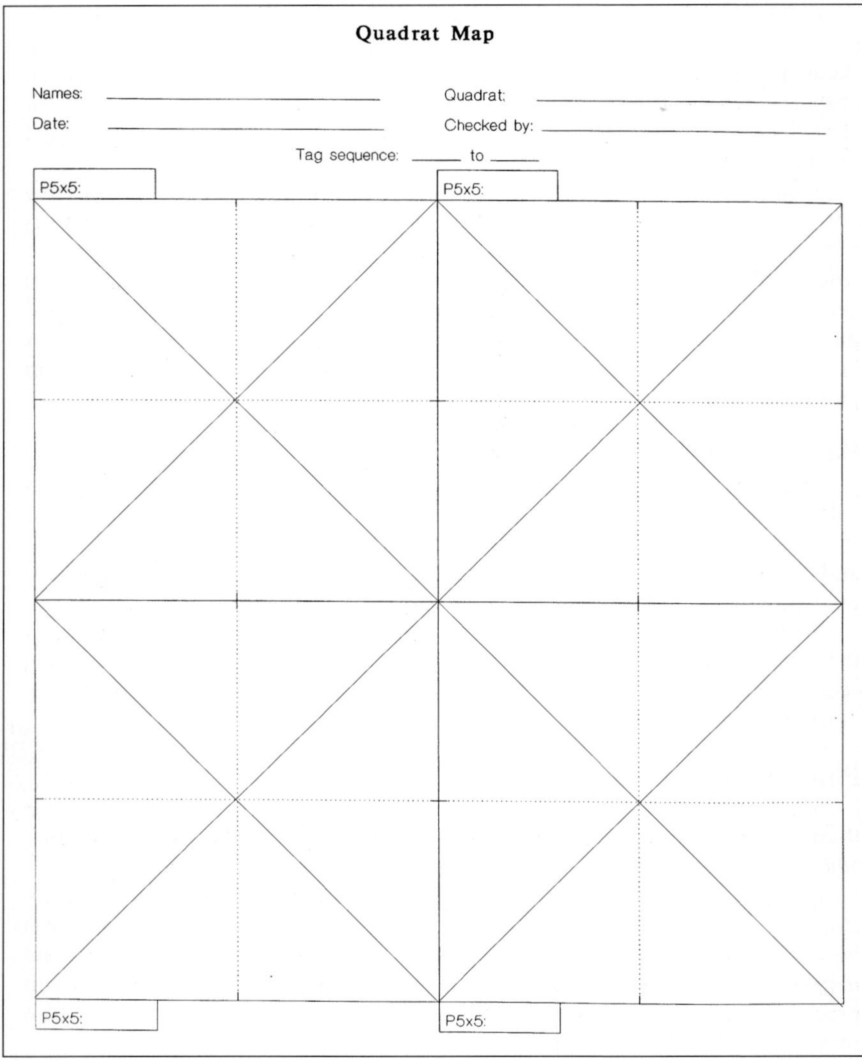

Quadrat Map

Names: ______ Quadrat: ______

Date: ______ Checked by: ______

Tag sequence: ____ to ____

P5x5: P5x5:

P5x5: P5x5:

Fig. 2.2.1. Data sheets. A) The field map of a 10x10 m region. (Figure continues for the next 5 pages.)

MAIN DATA

Quadrat: ____,____ No. Plants: ____

Name: ____________________ Date: ____________________

Checked by: ______________ Date: ______ 1. Data entry by: ______________

ID Check by: ______________ Date: ______ 2. Data entry by: ______________

Subquad (x,y)	Tag (6 dig)	Species (4 char)	Diam (mm)	Codes	Subquad (x,y)	Tag (6 dig)	Species (4 char)	Diam (mm)	Codes
___,___					___,___				

MD

Fig. 2.2.1B. The main sheet, where data from every tree are recorded.

MULTIPLE STEMS

Quadrat: ____, ____

Name: ______________________ Date: ______________________

Checked by: ______________ Date: ______ 1. Data entry by: ______________

ID Check by: ______________ Date: ______ 2. Data entry by: ______________

Subquad (x,y)	Tag (6 dig)	Diam (mm)	Diam (mm)	Diam (mm)	Subquad (x,y)	Tag (6 dig)	Diam (mm)	Diam (mm)	Diam (mm)

M

Fig. 2.2.1C. The multiple-stem sheet, for all individuals with code 'M' on the main sheet.

PROBLEMS

Quadrat: ____, ____

Name and Date: ________________

1. Checked by: ________________

2. Checked by: ________________

nq = does not qualify
lt = lost tag
id = identification problem
num = wrong tag number
two = plants with two tags
meas = measurement problem
dub = dubious tag or plant
alive? = possibly dead

Subquad (x,y)	Tag number	Problem code	Explanation of the problem

P

Fig. 2.2.1D. The problem sheet, for all individuals with code 'P' on the main sheet.

BUTTRESSED TREES

Name: ____________________ Date: ____________________

1. Data entry by: ____________________ 2. Data entry by: ____________________

(If the measure is a circumference measure, put in parentheses.)
(Enter "y" if the trunk is cylindrical or "n" if it is not.)

Quad (x,y)	Subquad (x,y)	Tag (6 dig)	Species (4 char)	First Height			Second Height		
				diam	ht(m)		**diam**	ht(m)	

B

Fig. 2.2.1E. The big-tree sheet, for all individuals with code 'B' on the main sheet.

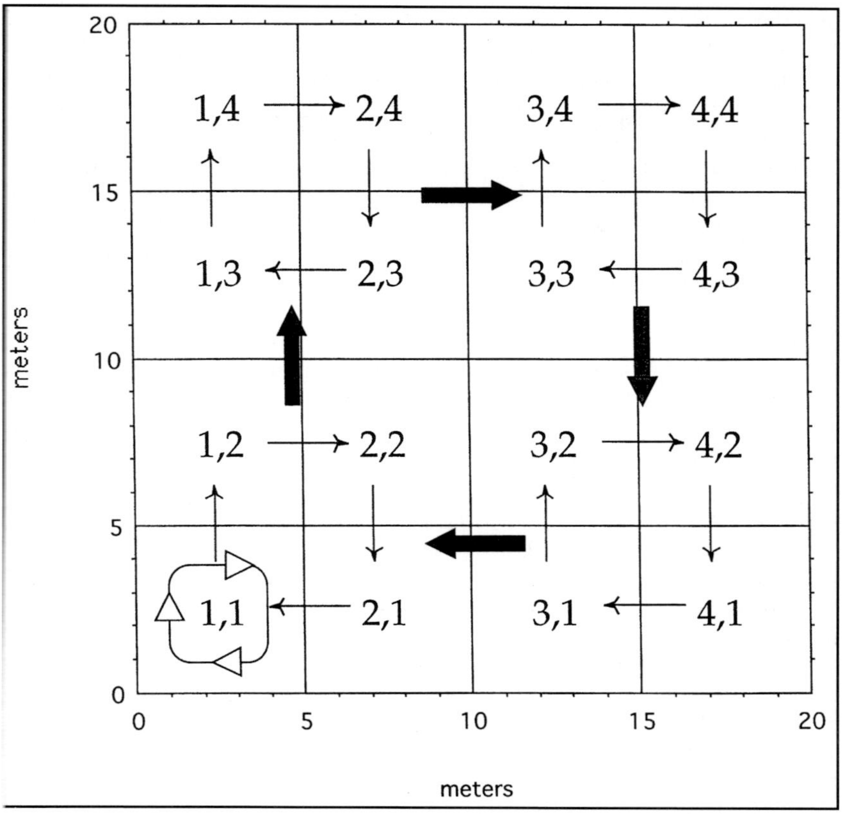

Fig. 2.2.2. Direction of mapping work within a 20 x 20 m quadrat. It always proceeds in a clockwise direction, within each subquadrat (shown in 1,1 only), within 10 x 10 m regions, and within the 20 x 20 m quadrat, as indicated by arrows.

added: one sheet should only have data from a single quadrat. On the other hand, when a quadrat is not finished at the end of one day, data from the same quadrat on the next day *should* go on the same sheets.

Aside from the four main types for each quadrat, there is a fifth field sheet which is used later (Fig. 2.2.1E). This is for big trees whose trunk diameters are measured separately because special equipment is necessary. Since there are usually two or fewer big trees per quadrat, one field sheet per column of data (25 quadrats) suffices. These sheets are not carried in the field during the main census.

Mapping Trees

Once a 5-m grid is marked, mapping the location of each tree is fairly easy, indeed, mapping to a precision of 5 m is already accomplished. Trees are mapped entirely by eye by marking a spot on the field map (Fig. 2.2.1A); large trunks should be indicated with large circles, the center determining the tree's location. No coordinates are entered in the field—coordinates will be established later when the field sheet is digitized. Trees are mapped where they are rooted, so leaning or prostrate trunks require care: the crown and most of the trunk may not be in the rooting subquadrat. If a prostrate tree has roots along a substantial length of the trunk, then the trunk base is mapped.

Each tree's tag number is recorded next to its spot on the map. Only the last three or four digits are necessary, with full tag numbers entered at the top of the map. For example, if a quadrat has tags 121627-121950, then these two numbers are recorded above the map, and only the final three digits (627 to 950) are written next to the trees. The next quadrat might include tags 121951-122207, so the last four digits would be recorded for each tree.

Tagging Trees

Tags are simply attached with loops of string around the trunks. Loops must be large enough to allow for growth: 600 mm for stems < 60 mm dbh, 1 m for trees < 150 mm dbh, and 2 m for trees < 300 mm dbh. The string is tied and laid on the ground, except on multiple-stemmed plants (see below). Larger trees (≥ 300 mm) are tagged with nails, with at least 50-60 mm of nail left above the bark. Nails should not be placed at the spot where the dbh is measured.

Tags should be placed in a numerical sequence that matches the work sequence: clockwise within quadrats, northward up columns, etc. If several teams work at once, a sequence of numbers must be assigned to each column. Conveniently, assigning 10,000 tags per column works out perfectly in most plots with 500-m long columns. This means that the first two digits on a tag correspond with the tree's column (e.g., column 00 has tags 000001-009999, column 01 has tags 010000-019999, etc.). Some tags from each column remain unused—perhaps they can be used in the next census or in another plot. If one column exhausts all 10,000—this should be rare—then numbers from a neighboring column can be used. (If tree density frequently exceeds 10,000 per column, then an alternate numbering system should be employed; see below.)

Tag sequences should be prepared before going into the field. The tags are strung in order onto a piece of stiff plastic tape large enough to hold the tags tightly, but allowing them to be pulled off in order (12 gauge vinyl tie-tape from Forestry Supplier works well). A sufficient number of 600-mm pieces of thread should be precut and carried so that single pieces can be easily pulled out. At BCI, we invented a simple device for cutting string into 600-mm sections: a roll of thread is placed flat on a board around a peg, so that it can be spun. A cutting blade is fastened to the board near the peg, and a marker placed on the board exactly 600 mm from the blade. One person can thus produce many pieces quickly. About 30 1-m pieces should also be cut for every quadrat (for trees 60-150 mm dbh), and a few 2-m pieces for larger trees.

Measuring Trees

Good measures of stem dbh are critical, since a key result from repeat censuses is growth. Recording the dbh is straightforward on regular, cylindrical stems: calipers are used for trees < 60 mm dbh, and a diameter tape for larger trees. In either case, the dbh must be measured exactly perpendicular to the trunk. Breast-height is found with a pole exactly 1.3 m in length, which is placed against the tree. Dbh is recorded to the nearest millimeter, but dbhs < 10 mm are not rounded up (9.9-mm trees are not included).

Unfortunately, stems are seldom perfect cylinders, and therefore quite a variety of difficulties in measuring dbh arise. To assure accurate, long-term records of growth, there is one fundamental adage to keep in mind: think about the next census! Consider whether the next field worker, five years in the future, will record a dbh that is exactly comparable. The following 10 rules of tree measurement are designed to ensure that he or she will.

Rule 1

When using calipers, take the largest dbh. Most stems are, in fact, not circular in cross section, and calipers will record different dbhs, depending on their orientation. With a diameter tape, this problem vanishes, but stems < 60 mm dbh cannot be measured accurately with tapes. Finding the largest dbh is easily accomplished by rotating the calipers while they are clamped lightly on the trunk.

Rule 2

Breast-height is always calculated on the uphill side. Breast-height on the downhill side would be lower.

Rule 3

Breast-height is taken along the lower side of a leaning tree, not the upper side (Fig. 2.2.3).

Rule 4

Breast-height includes all stem above the ground, no matter the angle. Thus, a leaning stem may be measured only a few centimeters above the ground (Fig. 2.2.3), and on a sharply curved stem, 1.3 m must be measured around the curves with a tape measure. However, in cases where a leaning stem has multiple rooting points so its origin is not clear, it should be treated specially, like a prostrate stem (rule 8).

Rule 5

Lianas, stranglers, and epiphyte roots should be pulled away from the trunk and the dbh tape slid underneath whenever possible. When the epiphytes cannot be moved, large wooden calipers with pointed tips can be used to avoid them (Fig. 2.2.4). Any tree requiring the large calipers should be marked as a big tree, to be measured later by the big-tree team (see rule 7).

The remaining five rules govern situations where something prevents measurement at breast-height, usually an irregularity in the stem or uncertainty about just where breast-height is. In these situations, the dbh must be taken at a different height, and thus is no longer a dbh, just a diameter. In the following discussion, I use the term "point-of-measure" (POM) to refer to the position at which a diameter is taken.

For all cases where the POM is not breast height (rules 6-10), these should be marked with paint so that they can be precisely re-located during the next census. A code indicating that an alternate POM was used must be recorded, and next to the code, the height of the POM is entered in parentheses (Table 2.2.1).

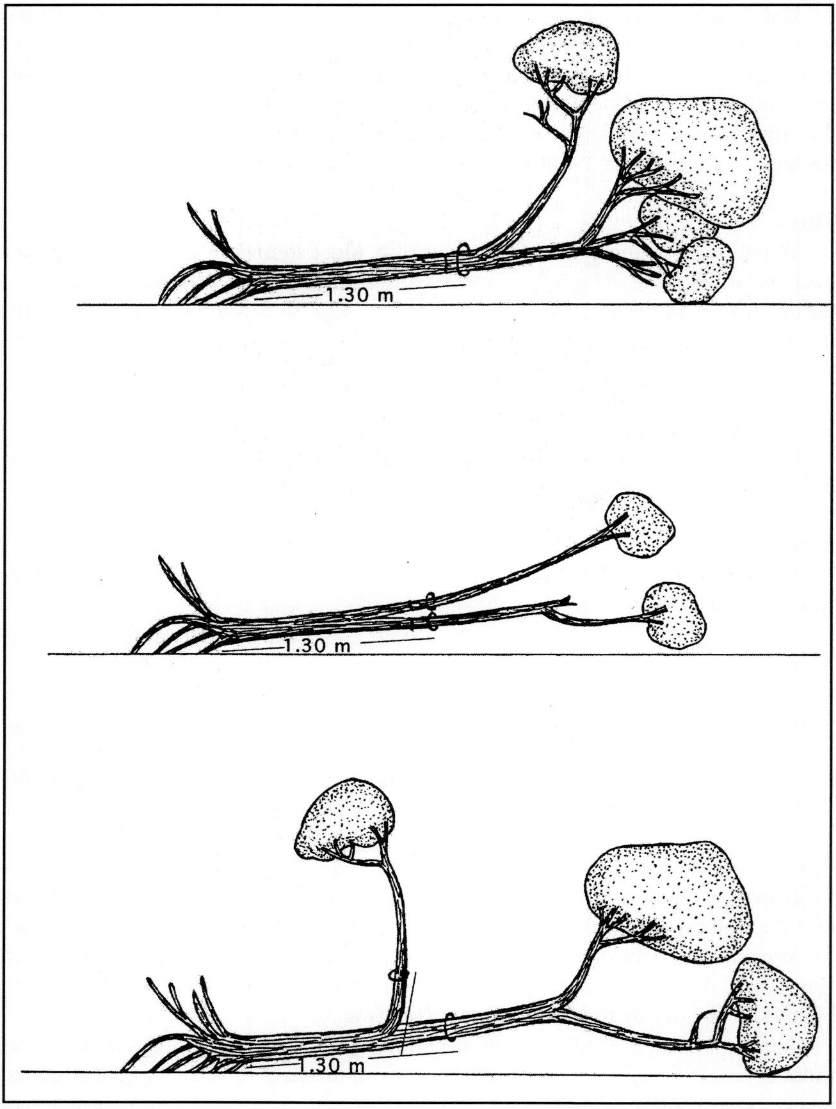

Fig. 2.2.3. Measurement of leaning trees. No matter what the angle, the dbh is taken 1.3 m from the base measured along the stem, following curves if necessary.

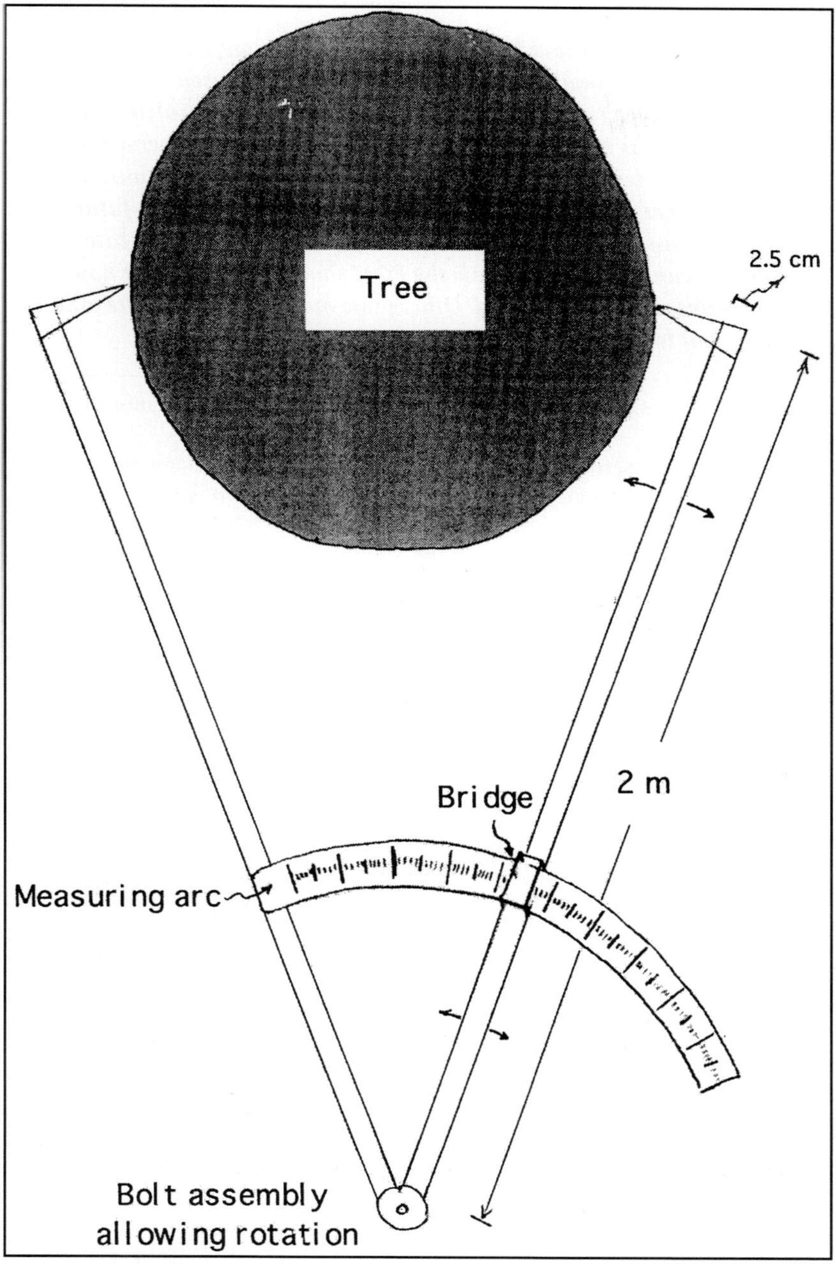

Fig. 2.2.4. Giant calipers for measuring big trunks hidden by stranglers. It is made of wood except for the bolt assembly holding the two arms together. The measuring scale can be drawn on the arc with waterproof pen by laying the tip of the calipers against a tape measure at various lengths. If the arc is one-third of the way out the arms, then the device has one-third the resolution of the measuring tape. It should only be used in unusual cases where the tape cannot be wrapped around the trunk.

Table 2.2.1. Codes that should be used during a first-time large plot census. All are entered on the main data form. The first two are used to indicate that a tree's dbh is entered on a different datasheet (big-tree or multiple-stem). The problem code is used to indicate any plant requiring further attention. The next three codes relate to the point-of-measure (POM) on unusual trunks, and are essential for accurate re-measurements during a future census (as discussed more fully in chapter 2.5). The last three indicate stem features that can help indicate how the POM should be located, or how multiple stems should be counted. Other plots may wish to institute other codes relevant to the local biology.

code	use	where recorded	additional information
B	large buttress, requiring ladder to measure	code column	dbh entered on big-tree sheet (later by big-tree team)
M	multiple stems	code column	dbh entered on multiple-stem sheet
P	any problem requiring further attention	problem column	description on problem sheet
A	POM at alternative height, not breast height	code column	height of POM also entered in code column
I	stem irregular where measured	code column	
X	stem broken below breast height	code column	
L	stem leaning	code column	
Y	stem prostrate	code column	height of POM also entered in code column
Q	stem broken above breast height	code column	

Rule 6

When an otherwise cylindrical stem has an obvious swelling or constriction at breast height (Fig. 2.2.5), the diameter is taken 20 mm below the lowest point of the irregularity. A thumb's width is used as an estimate of 20 mm.

Rule 7

For trees with buttresses or stilts (Fig. 2.2.5), the diameter must be measured at least 50 cm above the top end of the highest buttress (or stilt). If breast height is 50 cm above, then the measurement is made as usual. However, often it will be necessary to measure further up the trunk. If a mapper can reach 50 cm above the buttresses without a ladder, the measurement should be made immediately, and the

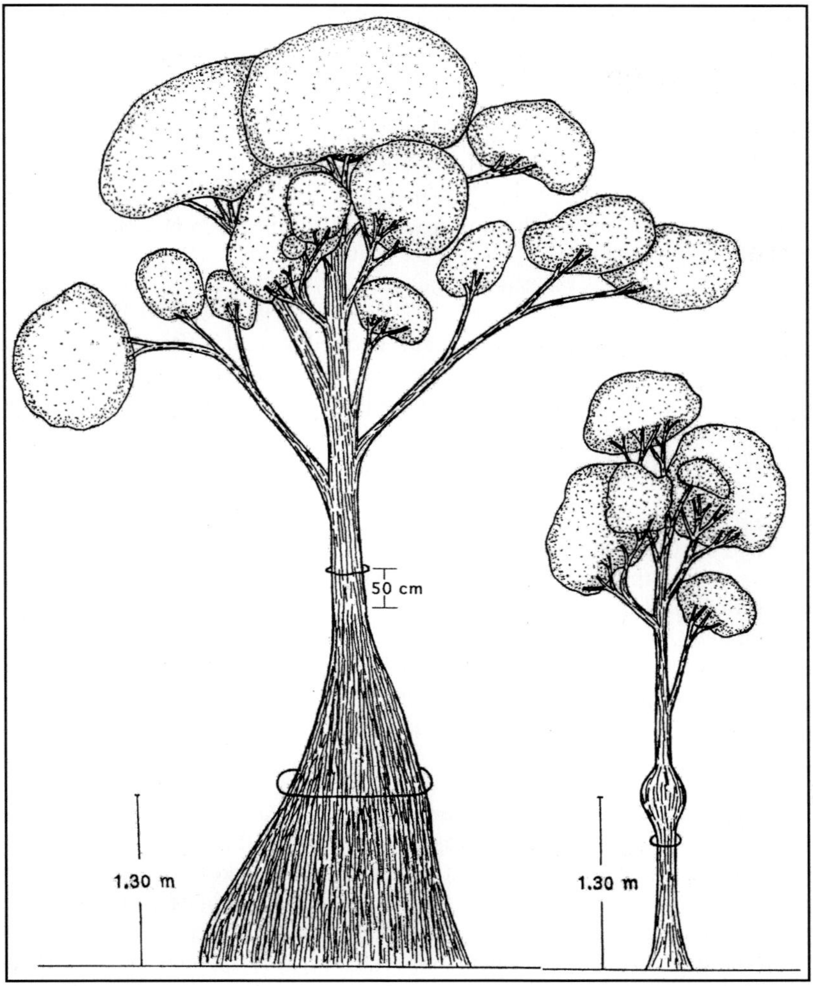

Fig. 2.2.5. Measurement of irregular trunks: 50 cm above buttresses, but 2 cm below swellings.

POM marked with paint. But for many buttressed trees, a ladder must be used to reach the POM. In this case, the original mapping team simply records 'B' in the code column for the tree and leaves the dbh blank.

After finishing a column, the big trees are measured all at once by the mapping team. (This can probably done with three people, so two team members would be available for other chores such as Xeroxing data sheets.) It is convenient, but not necessary, to wait until data for the column have been entered in the computer, so that all records with the code 'B' can be readily extracted and printed to a sheet (chapter 3.1). The big-tree team carries a ladder in order to hold the diameter tape above the buttresses. When buttresses are too high, then the measurement should be made by hoisting the tape with two wooden poles. The zero end of the tape is attached to one pole with a metal clip and a short thread, and the other end is walked around the tree and hoisted with a second pole. Binoculars are used to make the reading.

For many big trees, the diameter tape may not be long enough, so that a standard tape measure has to be used. If so, the reading is a circumference, not a diameter, and this should be carefully indicated on the data sheet. The big-tree teams should convert circumference to diameter immediately after returning to the field camp—leaving conversions until later risks confusion and some drastically oversized trees.

Rule 8

Prostrate stems should always be paint-marked, since it is often not clear where 1.3 m along the trunk is. In general, they can be treated like leaning stems, with the dbh taken 1.3 m along the stem from its origin (rule 4). Multiple-rooting points and buried trunks can cause confusion, however. If so, a point above ground should be selected and painted.

Rule 9

When a trunk is extremely irregular at all heights, a POM must be chosen as best as possible, then painted and recorded so the tree can be re-measured at the same spot in the future. Irregular trunks are the norm in some species, such as the highly fluted *Quararibea asterolepis* and *Platypodium elegans* at BCI, or the many cauliflorous *Cola* species in Cameroon which have protuberances on the trunk. A code indicating that the trunk was irregular at the POM should be recorded so that future mappers know.

Rule 10

This is for trees where no rule works: there is no rule 10. There will inevitably be some trunks that simply defy measure, for instance when completely buried in a strangler fig. The best possible measurement must be taken—at least some estimate is better than none—and the location of the POM painted and recorded. As long as these cases are rare, the best estimate of the team leader is acceptable. However, if there is a consistent problem unforeseen by rules 1-9, then a new technique must be developed and applied regularly.

Multiple-Stemmed Individuals

I have left out, thus far, cases where a single plant branches or forks below 1.3 m, or has more than one stem connected underground. This situation requires special rules for mapping, tagging, and measuring (Fig. 2.2.6).

Separate stems which are obviously connected to one another below breast height, either above or below ground, are considered part of the same individual. A single tag should be tied or nailed to the largest stem of the clone. It is very important to mark the largest stem in this way so it can be identified five years later during the recensus! I discuss alternate methods for doing so below.

The full extent of a plant's cluster is drawn on the map (but later, only the center point of the cluster will be digitized). Each stem $\geq$ 10 mm dbh is measured following the rules described above, and every one of the diameters is recorded on the multiple-stem data sheet (Fig. 2.2.1C). If one stem is irregular or buttressed, its POM may not be at breast height, but all other stems would still be measured at breast height.

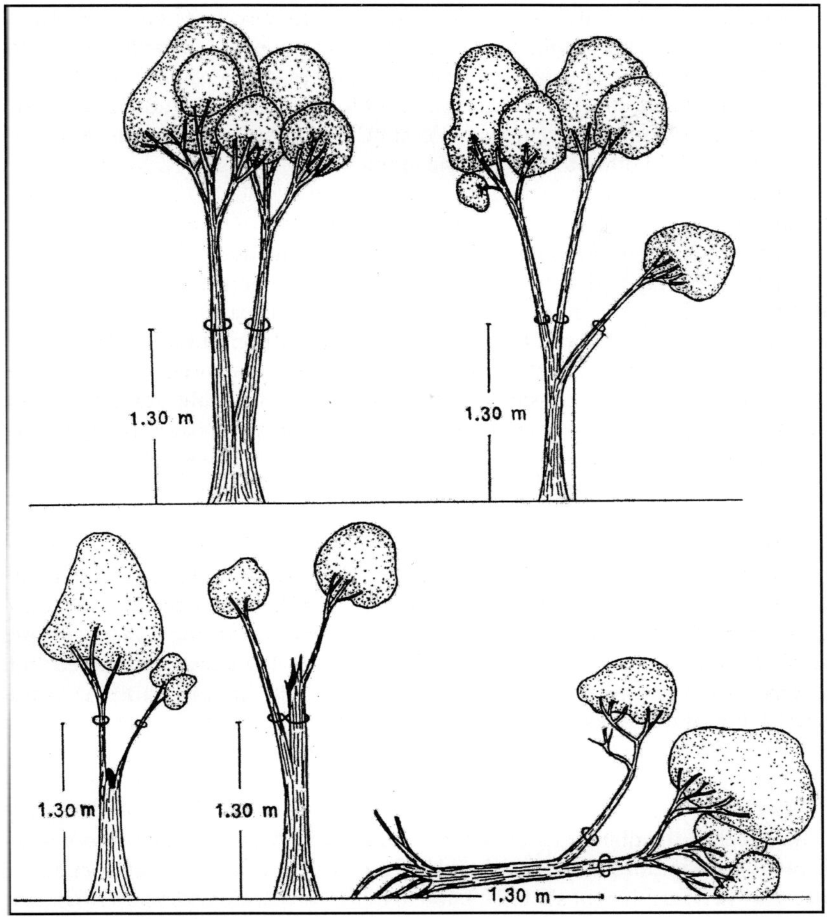

Fig. 2.2.6. Measurement of multiple stems. Any fork between the base and 1.3 m counts as a second stem, and must be measured 1.3 m from the base of the tree.

Any branch below breast height is considered a second stem, even if it is horizontal, as long as it is ≥ 10 mm dbh. Breast height is found by measuring from the ground, following any forks (Fig. 2.2.6). A branch above breast height is not considered a stem, even if the tree has buttresses so that its diameter was taken above the branch. If there are extra stems but they are < 10 mm dbh, the plant is not considered multiple.

Prostrate stems can have many vertical sprouts. At BCI, we only consider those within 1.3 m of the trunk base (where it roots) as stems (Fig. 2.2.3). However, if the prostrate stem roots at many points, then each vertical sprout should be measured 1.3 m above its base (i.e., above the prostrate trunk). Stems should be painted in this circumstance to prevent confusion (rule 8).

In some cases, it is not clear whether adjacent stems are connected underground. For species that do not normally grow as clones, a connection must be reasonably obvious to count two stems as one individual. Different rules should be developed for normally clonal species, such as species of the palm *Bactris* at BCI and Yasuní, or *Helicteres* at Mudumalai. For example, at BCI we consider any two *Bactris* stems within 1 m of each other to be the same individual.

Palms

In many palms, stems can be much fatter than 10 mm dbh but still not 1.3 m tall, and these are not included in the census. The top of the stem is defined as the base of the lowest living leaf sheath. A difficulty arises when dead leaf sheathes persist, hiding the base of the lowest living sheath, so the location of the top of the stem must be estimated. When measuring the dbh, dead sheathes are not removed, but are pressed as tightly as possible against the stem. Leaf petioles do not count as part of a stem even when they resemble a stem. Thus palms are often excluded even though their leaves are much taller than 1.3 m, and likewise for tree ferns or certain monocots with long petioles.

Stranglers

Stranglers are measured when the host tree is gone. Trunks are convoluted, and sometimes several trunks near the ground merge further up, so the POM of stranglers is problematic (and thus should always be painted). In some cases, a single trunk is measured well above the merging point, as if it were a buttressed tree; however, if there is no clear, single trunk at any height, separate trunks are measured at breast height as if they were multiple stems. In the latter case, the largest trunk should receive the tag, just as in multiple-stemmed plants.

Other Data

Other than the dbh and information about the POM, a few other pieces of information are important to record (Table 2.2.1). Stems which show signs of an old break below breast height should be noted; this information will be useful during the second census. Stems which are broken above breast height should also be noted, and stems leaning by more than 45° or prostrate should be indicated. Each of these pieces of information should be recorded by a single-letter code (Table 2.2.1).

Problems

Any plant that requires future attention is marked in the problem column on the main data sheet, and the nature of the problem explained on the problem data sheet (Fig. 2.2.1D). Problems include those situations where the rules seem not to apply and which prompt the supervisors to consult with each other or the chief scientists before deciding what to do. They also refer to simple things, such as lost tags, which cannot be corrected until the next day. When the problem is resolved, the entries should be checked off on both sheets to clearly indicate that no further attention is required.

Summary

By way of a summary of all the data collected by the mapping teams, I will review what is entered on each of the data sheets. The map sheet was already described in some detail (Fig. 2.2.1A), and includes a circle or a point for the location of each tree and the tree's tag number (the last three or four digits) written next to it. The range of tag numbers used in the quadrat should be written at the top of the sheet.

The main data sheet is where most other information about each individual is recorded (Fig. 2.2.1B). As for all sheets, the quadrat number, the first date a quadrat is censused, and the mappers' names are recorded at the top. For each plant, there are blanks for the following information: subquadrat number, tag number, species name, dbh, codes, and problems. Subquadrat number and tag number are straightforward. Size in millimeters is entered in the dbh column, except for multiple-stemmed plants or big trees, which get a blank dbh on the main data sheet. Species identification will be handled by separate taxonomy teams (chapter 2.3), but mappers should enter a species name if they know it.

After dbh come the codes, where a variety of information about a tree's status can be recorded. At a minimum, codes should be used for the situations described in Table 2.2.1. The codes column can also be used to indicate other features of a tree, depending on what is considered important at a site. The final column of the main sheet is used simply to indicate problems which require further attention.

The main data sheet can handle only one dbh per plant, so there is a separate data sheet for multiple-stemmed plants (Fig. 2.2.1C). Each tree getting an 'M' in its code column on the main sheet must be listed on this data sheet. Blanks for the tag and subquadrat numbers are included, along with three blanks for dbhs. Trees with many more stems (up to 105 on BCI) are entered on several lines. Each map team has one of these sheets for each quadrat.

The last data sheet is for big trees, and it includes data from one entire column of quadrats (Fig. 2.2.1E). Quadrat, subquadrat, and tag number are entered for each tree, plus the dbh and the height at which the dbh was recorded. The date is important for big trees, because they are often measured some weeks after the main work in a quadrat. A single big-tree data sheet should only be used over a span of two weeks, then all the trees on it can be given the same census date. This greatly simplifies database construction (chapter 3.2).

Phase 3—Checking the Work

An advantage of working in teams, with field supervisors recording data, is that all work can be checked immediately (in contrast, if mappers work alone, the supervisors should visit each quadrat afterwards). While recording data, each supervisor should also be monitoring the measurers and mappers. Occasionally, especially in difficult cases, the supervisor should re-measure a dbh to check accuracy. This is especially important early in the census, when measurers are inexperienced.

In addition, each of the team members should be encouraged to check each other's work. Both the person drawing the map and the one tying the tags can look for trees and check whether any have been missed. Anyone in the group might also notice that a dbh called out seems way off the mark, prompting a re-measure.

Finally, after returning to the field office at the end of each day, supervisors should double-check the data sheets for the current day. Are all sheets present? Does every record have a complete set of legible data? Are there duplicate tag numbers? Do all trees with a code 'M' on the main data sheet appear on the multiple-stem sheet, and vice versa? Problem sheets should be reviewed and problems resolved immediately, when possible.

If the quadrat was not finished, the data sheets are set aside, to be used the next day. If the quadrat was completed, the full set should be placed in a manila folder, with one manila folder per quadrat. Problem sheets, though, should be kept in a separate problem folder until they are fully resolved. The folders should be kept in a file cabinet at the field camp until the taxonomy teams need them.

Rationale for the Methods and Alternative Approaches

Each of the 12 large CTFS plots has arrived at somewhat different methods for mapping, tagging and measuring. Most of these differences reflect preferences about collecting exactly the same information and do not affect the comparability of the data sets. However, they may affect efficiency or cost. I have outlined the methods that I believe best optimize the trade-off between quality of data and rate of progress, but other field workers may disagree, and I outline here some of those disagreements.

There are some differences among the plots that *do* affect the data and thus the comparability of the data sets. Various plots have differed as to which species are included in the census, and I discuss these differences in some detail. There are also inevitable differences in how the point-of-measure is selected in unusual trees, and I mention some here.

Field Teams

All plots in Asia and Africa are mapped in large teams. Teams of two are used in Malaysia, but teams of 5-12 have been employed in the Congo, Cameroon, and Sri Lanka. The Luquillo plot uses teams of three. This is obviously a matter of preference, but it may also be a matter of efficiency.

In Ecuador, the census began with each mapper working alone, but we subsequently switched to teams of five. The rate of progress was identical either way—50-60 trees mapped per person per day. But the workers clearly prefer working in a group, and with teams, checking work is reduced (chapter 2.4).

Plants to Include

The rules for plants to include that I outline above are exactly the rules that have been followed at BCI. Other plots have allowed some variations, though, which fall into two categories. First are rules which deliberately include lianas or epiphytic trees; second are rules for plants which fall into a gray area regarding the definition of a free-standing, woody plant.

At BCI, lianas were excluded from the beginning for strictly practical reasons—they were thought to require a different set of census rules which would pose added difficulties. The main concern is that liana stems can be extremely convoluted and have multiple rooting points. Most other plots have followed this lead and omitted lianas, but two have not—Sinharaja in Sri Lanka and Ituri in the Congo—but the two sites have not followed a standard protocol. Because they are not standard, I describe the liana censuses separately (chapter 2.6).

At Yasuní, a different sort of problem has arisen with lianas because we do not yet know the flora well. We decided that all free-standing juveniles should be tagged and censused, even when suspected to be lianas, simply because we do not know for certain in all cases. When we find out, we will indicate each liana with a code in the database, but leave the plant tagged. Lianas relying on a host for support are not mapped.

The remaining rule differences regarding plants to include pertain to species which defy easy classification as either free-standing or woody. Strangler figs are an obvious case-in-point, and different plots have adopted different rules. They are measured only when free-standing at BCI (as described above), but were completely omitted at Pasoh. The Luquillo plot deliberately includes all stranglers that are rooted, even if still supported by the host, and at Yasuní, stranglers with at least one trunk ≥ 200 mm dbh are included, regardless of the status of the host. In the plot in montane forest in Colombia, all epiphytic trees (many *Clusea* in the Guttiferae) are included in the census, since they form an abundant component of the canopy.

Palms are also a source of some variation in rules. At most plots, plants are included when the lowest living leaf attaches to the stem above 1.3 m, but at Luquillo, they are included when the highest (newest) leaf arises 1.3 m from the ground. The diameter is measured around the lowest leaf base, or at 1.3 m, whichever is lower. Therefore more palms are included and the POM is lower, however, the difference is not great.

Other confusing species are bamboos and large herbs. A single bamboo species was mapped at Mudumalai, and is quite numerous, but bamboos at Lambir were excluded. Although at BCI we omitted all herbaceous monocots, large Musaceae were censused at Luquillo when taller than 50 cm. An abundant species of *Desmodium* that is usually a small herb at Mudumalai was excluded from the census even though it occasionally has upright, woody stems. At Sinharaja, a similar situation arose with *Strobilanthes* spp.; these were not censused individually, but cover estimates for each subquadrat were recorded.

These differences must be considered when comparing data across plots. Regarding the lianas and stranglers, data can still be compared as long as each is indicated as such with a code. Then it is possible to extract the portion of the data set that includes free-standing trees and compare it with another plot's data. The other between-plot differences govern only a small number of species, so the data from all the large plots remain fundamentally comparable. Table 2.7.1 in chapter 2.7 summarizes the variation in what individuals are included.

Work Sequence

During the first census at BCI, all big trees were tagged and measured separately, then smaller stems were added on a second pass. Moreover, the census was done "backwards", starting in the northeast corner and proceeding south and west. The direction is obviously arbitrary, what matters is that it be followed consistently. Unfortunately, it was not. Later censuses at BCI were carried out from southwest towards northeast, reversing the order.

At Luquillo, larger trees (≥ 100 mm dbh) were also mapped first, followed by a census of the smaller stems. This was done because the first goal of the project was to assess the impact of Hurricane Hugo, which had just passed, and small stems would have slowed the census (Zimmerman et al 1994). In general, though, it is better to census all trees together, so the census date is consistent across size classes.

Tree Mapping

During the original census at BCI, each map sheet covered an entire 20x20 m quadrat, but was printed on oversized paper. Large clipboards were used for writing. Although larger maps add accuracy, they are much harder to handle and photocopy.

Demarcating a subquadrat with rope while mapping is a convenience that has been done differently at different sites. In the Congo, subquadrats were outlined with four tape measures, and no 5x5 m stakes were ever placed. Trees were mapped by measuring the perpendicular distance from two boundary tapes. These numbers were then used to mark the tree's location on the field map, which was a piece of graph paper (so coordinates could be accurately located). This process made it possible to map an entire 20x20 m on a single, standard-sized graph paper. A digitizer was used later to enter the coordinates in a computer. A similar approach was used at Pasoh (Manokaran et al 1990). At Yasuní, diagonal strings were placed across each subquadrat but not around the perimeter, and plants were mapped by eye.

At Luquillo, trees < 100 mm dbh were not mapped, but their 5 x 5 m subquadrat was recorded. Thus, their position is known to only 5-m precision. Larger trees were mapped exactly.

Use of a field map can be completely omitted if coordinates are measured directly. As at Ituri or Pasoh, perpendicular distances to subquadrat boundaries can be measured, or distances to two corner stakes can be measured and later converted to perpendicular distances (Dallmeier et al 1991). Without use of a map, more work is necessary in the field, but plants' coordinates need never be digitized.

There are two reasons I recommend using a field map and placing trees on it by eye. First, a field map allows other features to be drawn, such as fallen trunks, rocks, streams, or animal nests, and these can be added to the geographical database (Manokaran et al 1990). Second, a field map saves time in the field but adds office work, and I argue that this is generally a good thing. Field work is more difficult, more subject to error, and more limited by weather than office work, so shifting labor from field to office should heighten efficiency.

Tags and Tagging

The size and shape of tags is unimportant, and it may be desirable at some sites to produce tags locally. Large sheets of aluminum are cut to the appropriate size, and numbers punched by hand in each one. Tags for the Korup plot in Cameroon were produced like this at considerable savings, but this depends on local wages.

Various approaches for attaching tags to trees are in use in large plots. At Barro Colorado, we have always used tough, green vinyl tie-tape from Forestry Supplier (12 gauge [0.2 mm thick]x13 mm wide, catalog #79317), however, the green is attractive to leaf-cutters ants (*Atta* spp.) and cannot be used where the ants are numerous. At Yasuní, in Ecuador, we switched from green tape to thread because ants were a problem. Primates and other mammals are also sometimes attracted by the plastic tape.

At Lambir, a long copper wire holding the tag is twisted so that it hangs at eye level on the trunk, but has plenty of length to allow growth. Wires can also be placed on the ground. Wires probably last longer than plastic tape: the tape in use at BCI becomes brittle and easily broken after 5-7 years; thread has persisted well for >10 years on trees in Cameroon, but I have no information on wire.

In most plots, nails have been used to place tags in trees larger than 80 mm dbh. The problem is that nails must be long enough to allow growth, or the trees "eat" the nail. Substantial numbers of trees add 50 mm of radius in 5 years at BCI, and a few grow even faster. Since thread is cheap and easy-to-tie, I see no reason to use nails on trees < 300 mm dbh. At Yasuní, no nails are being used at all, even on the largest trees. On the other hand, I have no direct evidence that tag scars damage trees.

Tag numbering schemes at different plots have been highly variable. That each tree has a unique number is all that really matters, but there are two conveniences to the use of logical tag sequences. First, it makes it easier to assign tags to different field teams. Second, it is useful to future workers searching for a particular tree to know that tag numbers reflect column numbers and that nearby plants have numerically close tag numbers. One logical and compelling alternative to the one I outlined above is to use tag numbers with two sections: a quadrat number and a tree number: trees in quadrat 3513 receive tags 3513-000, 3513-001, 3513-002 ... 3513-177, 3513-178, etc. Or, tags can have a 2-digit column number and a 4-digit tree number. The Lambir plot employs the quadrat approach, and the Korup plot uses the column approach.

At the BCI plot, two numerical sequences of tags were applied during the first census, because large trees were mapped first and smaller ones subsequently. Neither sequence bore any relation to quadrat numbers, but nearby trees did receive nearby tag numbers. A similar approach to tagging was used at Luquillo.

Measuring the Trees

All plots now measure trees to the nearest millimeter (accuracy of the measurements will be treated in chapter 2.4 on checking the census). An alternative method described by Manokaran et al (1990) was formerly in use at BCI and Pasoh, though. A plastic plate holding a series of notches of 10, 15, 20, 25 ... 45, 50, 55 mm wide was used, and the largest notch size that did not fit around a stem was recorded. For example, if the 25-mm notch fit around a stem but not the 20-mm notch, the dbh was recorded as 20 mm. This produces dbhs rounded down to 5-mm intervals. The plates were used for all stems < 55 mm dbh at BCI in the first and second censuses and at Pasoh in the first census. Larger stems were measured with dbh tape to the nearest millimeter. The logic to this old approach was that the only growth that matters in the long-run is growth from sapling to adulthood, and thus the difference of 2-3 mm in saplings is trivial. We decided, however, that it was no more difficult to use calipers and record millimeter accuracy.

Regarding calipers and non-cylindrical stems, I recommend recording just the largest dbh. Unfortunately, using the large dbh means that a tree loses a little size when it grows from its caliper stage to its tape stage, and that basal area is overestimated slightly (Sheil 1995). The more precise alternative would be to record two dbhs per tree (maximum and minimum), but this would enlarge the database and add confusion. We have decided at BCI that it is not worth the added effort, and all plots are following this approach.

When stems are swollen, prior BCI rules called for measuring "just below the swelling". I now recommend a more precise rule, "20 mm below the swelling". Other plots have had minor differences in handling swellings. At Yasuní, measurements are taken at the smallest diameter between 1.3 m and the ground if 1.3 m is unavailable. At Luquillo, measurements were taken above or below 1.3 m, whichever was convenient (since the POM is painted at Luquillo, this does not cause inconsistencies).

Leaning and prostrate trunks are a source of variation in measuring rules. At Luquillo, vertical stems on a prostrate trunk—no matter how far from the rooting spot—are measured, and their diameters taken 1.3 vertical meters above the ground. Since Luquillo has many fallen trees that re-root and re-sprout (due to hurricane damage), adding all vertical sprouts is sensible. Some plots measure 1.3 m vertically from the ground to find the POM in all leaning stems, rather than along the trunk as we do at BCI. The problem with this is that when a tree falls, or leans, its POM changes. Fortunately, leaning and prostrate trunks are usually not common, so slight variation in how they are handled should not affect the overall comparability of data sets.

Buttressed trees often pose measurement problems (Sheil 1995; Clark and Clark 1996), and early methods at BCI were flawed so that growth data on large trees were not adequate through 1990 (Condit et al 1995b). But big trees are too important for carelessness because they hold the lion's share of the forest's biomass. The rule of staying 50 cm above buttresses, and the careful use of diameter tape on all but the most inaccessible trunks, should save other plots the mistakes we made at BCI and produce data on growth and biomass accumulation in even the largest individuals.

Our first mistake at BCI was to measure around buttresses, at breast height, regardless of where or how large the buttresses were. But diameters around buttresses make no sense, as they do not reflect bole size, and starting in 1985, we have measured just above buttresses. Unfortunately, buttresses grow up trees, and measurements should not be taken "just above buttresses", but rather, well above buttresses. Thus, we now use the rule "50 cm above the top buttress". I estimate rates of buttress ascension at < 3 cm per year, since trees I know to be < 100 years old always have buttresses < 2 m high, and giant buttresses—up to 11 m in one *Ceiba pentandra* at BCI—only occur in giant trees that must be > 300 years old. Thus, after 5 years, 50 cm of leeway means that the original POM would still be above the buttress.

Discussions of POM raise the important issue of whether trees should be painted where they are measured. At BCI, trees were never painted in 1982 nor 1985, and then only buttressed trees were painted in 1990 and 1995. At Lambir, Sinharaja, Ituri, and Luquillo, every single tree has been painted at the POM, and at Pasoh, all trees > 50 mm dbh were painted. At Yasuní, all trees ≥ 300 mm dbh are paint-marked, and all trees ≥ 100 mm if the POM is not breast height. Manokaran et al (1990) found that painting reduced the fraction of large dbh mismeasures (> 5 mm) from 2.7% to 1%, but so far, no growth data have been published from a large plot where every tree was painted. For now, I suggest that painting every POM is unnecessary, as the growth data at BCI have been adequate for most analyses, but that paint should be used in all cases where the POM differs from 1.3 m. Several plot managers feel very strongly that painting every stem is useful, though.

Even if trees are painted, the height of measure should be recorded when it is not 1.3 m. This is a precaution against the paint peeling away. At BCI through 1995, though, we only recorded the height in buttressed trees, not smaller trees. In future censuses, however, we will record all alternate heights.

Multiple Stems

These have posed a serious problem in the BCI census, because we have never (through 1995) placed tags or other marks to identify individual stems. It has thus been impossible to know which dbh goes with which stem. After several recensuses, I have learned the pitfall to this approach: if some stems of a multiple-stemmed plant die, it is impossible to judge the growth of the surviving stems. For example, consider a tree with 45-, 35-, and 25-mm stems in one census, then 46- and 37-mm stems in the next census. How much did the two stems grow? Perhaps it was the large stem that died, while the 35-mm stem grew to 46 mm, and the small one to 37 mm. Or perhaps the small one died, the large one grew 1 mm, etc. Without tags, there is no way of knowing.

The simplest way around this problem is to mark the largest stem by placing the tag there. Then the one major stem can always be identified and its growth monitored. Other marking methods would suffice, such as an additional loop of thread. At the Yasuní plot, a paint mark is placed on the largest stem, distinctive from the paint mark used for unusual POMs. These methods do not allow growth of secondary stems to be estimated, but growth of smaller stems is seldom used anyway. Since all stems are measured, basal area growth of the whole clone can always be estimated.

Alternatively, every stem from multiple-stemmed plants can be marked with an individual tag. This is done in the Mudumalai and Luquillo plots, where several species typically grow in large clusters. Small tags with letters (A, B, C...) can be used for individual stems. Tagging every multiple stem is a lot of work, however. A compromise has been adopted in the Ituri plot: any stem which differs by < 10 mm dbh from another stem on the same plant is tagged. This should clear up much potential confusion, however, I would recommend a larger limit, perhaps 30 mm.

Other Data

The codes listed in Table 2.2.1 represent the minimum information that ought to be recorded during the first census of a large plot. Many other pieces of data might be added, depending on what specific questions are important at a given site. For example, bole form is important in studies of timber production, so this might be useful to record. There are obviously many other features that might be ecologically or silviculturally interesting, but it should be kept in mind that the main goal of a large plot is to mark a very large number of trees, so recording lots of data about every stem will only slow down the work. Our view at BCI is that we collect the minimum information necessary to do basic demography, that is, to calculate population densities and growth and mortality rates. More detailed studies should be done separately from the main census, on a subset of the data, when specific questions arise.

Census Date

The census date is an important number for later analyses of growth or mortality, and thus must be recorded on every quadrat's data sheet. A single date is sufficient for each quadrat—the day on which work began in that quadrat—even though individual quadrats might take several days to finish. It is simply understood that a plant's census date is accurate to within two weeks, and given a 5-year interval until the next census, this is plenty accurate. For big trees measured weeks later, separate census dates must be recorded, and these will go in the final database (chapter 3.2). The obvious alternative is to record a date for every single tree, but this is hardly likely to be worth the trouble.

Time, Labor, and Costs

Supplies

Each field team needs calipers, diameter tapes, compasses, and a clipboard, all of which run about \$200 per team. For data sheets in a first-time census, about 28,000 pieces of standard-sized paper are needed—four maps per quadrat, five main sheets, one multiple-stem sheet, and one problem sheet. All data sheets need to be Xeroxed after the checking stage is finished (chapter 2.4). We have found that waterproof paper is useful but not essential for the work. It costs about 9 cents per page, compared to 0.9 cents for regular paper.

PVC stakes cost about \$1.20 each, thus for the 20x20 m corners of a 50 ha plot, about \$1,700 is needed. Adding permanent 5-m stakes would add at least \$7,000 more. Using stone markers is the only reasonable way to save substantially, since wooden stakes do not work.

The numbered tags from Forestry Supplier cost 5 cents each, but locally produced aluminum tags in Cameroon cost only <2 cents each. Polyethylene string cost 0.6 cents for each 60-mm section in Cameroon, after separating three threads per unit. For 300,000 trees then, the cost of tagging will be between \$4000 and \$18,000.

Other than these main items, the tree census will require a tag-punching machine, binoculars, a ladder, some rope, and several tape measures. This can cost a total of \$1500-\$2000.

Time and Labor

At Yasuní in Ecuador, mappers working solo or in teams can cover an average of 50-60 plants per person per day, neither identifying nor collecting plants. Precise records for BCI during the first census are not available, but it appears that the rate was similar based on the total amount of time it took to finish.

In Sri Lanka, 8-12 people worked as a team. One team mapped an average of 350 trees per day, or 35 plants per person per day, but these groups identified and collected as well. In Puerto Rico, teams of three covered 50-70 new plants per person per day. In the Congo, mappers worked in groups of eight and covered 80 plants per person per day, including identification and collection. The Congo teams worked longer days than elsewhere, though, and only worked 14 days per month. In Cameroon, teams of five covered 50-80 plants per person per day, working only 7 hours per day and not collecting plants.

Assuming 60 plants per person per day, and with 250,000-350,000 plants in a large plot, it would take 4000-6000 person-days to map the trees. Assuming 21 days a month and 10 workers, mapping would take 20-28 months. This does not include species identification and data checking, which are covered in the next two chapters.

Key Issues for the Tree Census

- the cardinal rule—think about the next census and how easily a plant can be re-located and re-measured
- trees are mapped by eye on blank maps of each 10x10 m region
- tags should be placed in numerical sequence so that trees within a quadrat have similar numbers
- dbh is easy to measure as long as trunks are straight and regular in shape
- special rules for handling leaning, prostrate, multiple, or irregular stems need to be carefully spelled out and should match a standardized approach
- buttressed trunks pose the greatest difficulty, and should always be measured 50 cm above the top buttress
- POMs at unusual heights should be indicated with paint, with the height recorded
- the largest stem of multiple-stemmed plants must be indicated by placing the tag there, or by a paint mark
- palms, tree-ferns, stranglers, and large herbaceous plants require special rules.

Chapter 2.3

Taxonomy

Identifying every one of a quarter million tropical plants to species is the most difficult aspect of finishing a large census. Because the vast majority of individual trees in a forest do not have reproductive structures at any given moment, and most are juveniles which will never reproduce, plants must be identified on the basis of sterile characteristics. Individuals must be recognized from leaves, stipules, branch form, odor, sap color, bark, etc. This requires experienced botanists.

On the other hand, most of the primary results of a large census do not require that specimens be named, only that they be sorted accurately into "morphospecies"—morphologically equivalent groups which are assumed to be single species. Eventually we must get full names, often by sending specimens to experts on given families, but most plot inventories include morphospecies which remain unnamed for years (usually known to genus, though). Nevertheless, ecological work proceeds.

My purpose in this section is to recommend a general protocol for identifying morphospecies in a large plot. It is not a taxonomic key, and I do not offer details about how morphospecies are separated. Instead, I focus on the logistics of collecting and sorting specimens in an area where the flora is poorly known. I will then briefly comment on plots where the flora is already well-known and the taxonomic work much easier. Finally, I will review alternative approaches taken at different plots, including those mistakes made during early censuses at BCI that have led to my current recommendations.

Personnel

Teams specializing in the taxonomy should be responsible for identifying every specimen. The mapping crews cannot: they must concentrate on the mapping work and do not have time to study the flora intensely. The mappers can be trained to recognize common species, and should fill in the species name on the main data sheet when they know it, however, they should not be relied upon for final identification.

Instead, an expert taxonomist for the area should train two or three promising botanists, and these trainees should work full-time on plant identification. They must be good at botany, but most important, they must be committed to doing the painstaking work necessary to accurately categorize thousands and thousands of fairly similar-looking leaves. The work requires patience and a virtual obsession with being right, so the trainees must be selected carefully. One way to select the botanists is from among the mappers, a few weeks after work begins, when their skills begin to show.

Tropical Forest Census Plots: Methods and Results from Barro Colorado Island, Panama and a Comparison with Other Plots, by Richard Condit.

After three or four months of training in the first two hectares of the large plot (see Phase 2 below), the taxonomy teams take on the full census. Each team is comprised of a taxonomist, an assistant for help in collecting and recording, and—if at all possible—a tree-climber. If there are two taxonomic teams, one tree-climber can suffice for both, as the two teams can work in close proximity. The assistants can be mappers; indeed, each of the mappers can take a turn as taxonomic assistant to learn more about the process. The taxonomic teams work behind the mapping teams, so plants are already mapped and tagged. Ideally, they start after data have been entered in the computer, to facilitate construction of a taxonomic database.

Phase 1—Review What Is Known

A review of existing collections from the region, at as many herbaria as possible, is a good starting point. If a checklist of the region's flora does not exist, it should be created by listing all tree and shrub species known from the area in a computer database. For each species, the family must be included in the list.

Phase 2—The First Two Hectares

The taxonomists' first goal should be to complete all identification work in two hectares within the large plot. Once the trees in these hectares have been tagged by the mapping team, the taxonomists begin. These two hectares serve as training, and all the new taxonomists, plus the chief taxonomist, should work together.

Collecting

Except for very well known species where it is certain that there will be no confusion, leaf specimens should be collected from every tagged plant in the two hectares. Portions of branchlets must be included when possible, and compound leaves should be collected in full. Flowers or fruits should always be taken when present. However, samples should not be taken from very small individuals where it might cause severe damage.

If tree-climbers are available, medium and large trees may not be a problem. Otherwise, extendible pruning shears can be used to get branches from large saplings and pole-sized trees, while leaves are collected from the ground beneath larger individuals. This requires carefully studying leaves on the tree through good binoculars, then finding matching leaves on the ground. With some practice, and careful study of leaf shape and venation, leaves can be collected in this manner for nearly every big tree. In the end, though, there will always be a few plants that cannot be collected; these will be visited later.

It is crucial during collection that tag numbers be carefully and accurately recorded. They should be written on tape that is wrapped around the corresponding branchlet, so that the specimen can never be separated from the number.

Field Notes

A small, waterproof notebook is useful for field notes, and the tag numbers for a given quadrat can be filled in before going in the field. Brief notes should be taken on every plant, especially emphasizing those characters that cannot be collected—bark, latex, trunk shape, branching pattern, buttresses, and roots. The team should aim to reduce these notes to just a few words, using abbreviations and codes.

A best guess for family and genus should be included with the field notes. It is thus crucial that the chief taxonomist teach family-level characters.

A Database

The taxonomists create their own computer database as work proceeds. This is necessary so that return visits to plants of certain families or genera can be easily arranged. The best way to do so is by starting with a copy of the main database for the quadrats under study (see chapter 3.1). Five pieces of data already in the main database—tag number, quadrat number, subquadrat number, codes, and species name—are kept, and several new fields necessary for the taxonomy work are added: one to indicate whether a plant has been collected, another for prior names so that name-changes can be documented, and a third for brief field notes. Other data-columns in the main database are not immediately relevant to the taxonomy and can be deleted (chapter 3.1).

The field notes should be entered in the database every day, for every plant, and the collection code added for each individual collected. As provisional species names are invented, these are entered as well.

Storing Specimens

The specimens should be stored at the field camp. They first must be dried and pressed, using either electric heaters or wood stoves if there is no electricity. Metal or wooden cabinets should be purchased or constructed to hold the specimens, as these can help protect from insects. If the field herbarium can be air-conditioned, this will help protect the specimens as well. If insects get in, one simple way to eliminate them is by freezing the collection for a couple days, but this will not always be feasible.

Sorting Morphospecies

After the first 20 x 20 m quadrat has been collected, the specimens should be sorted. A large table next to the storage cabinets is ideal. Specimens are first divided by family, and within by genus where possible. All morphologically distinct forms are then segregated, a process which relies entirely on the chief taxonomist's expertise: he or she must decide what morphological differences qualify as distinct species. Field notes are consulted while sorting so that bark or trunk characteristics can be considered as well. Obviously there will be some mistakes at this stage, and all sorting is provisional. Specimens are then given family labels and replaced in the cabinets alphabetically by family.

After about five 20x20 m quadrats are completed, or 1000-1500 plants, sorting should be repeated. All individuals of a given family should be placed on the table and studied together. This time, as species are separated, provisional names should be invented for each recognized morphospecies, and these should be entered in the database. Names that serve as mnemonics—"Lauraceae small-skinny", "Eugenia long-hairy", "Inga winged-square-6"—can be used. Of course, if the real Latin name is known, it should be used. The provisional names are entered in the taxonomy database.

Some questions will remain: some variability will not fall into clear groups, and a particularly vexing problem will be matching adults and juveniles, which often have different leaf forms. Thus, the species name should be left blank for many individuals. Later, as more specimens accumulate, the variants will sort themselves into distinct groups, and adults and juveniles will become easier to match as intermediate sizes are found.

Sorting should be repeated after subsequent blocks of quadrats are finished. As the taxonomists gain familiarity with the flora, sorting will be much easier and quicker, so it does not need to be done as frequently. Perhaps it will be necessary after 10 quadrats are finished, then not again until a full hectare (25 quadrats), and again after two hectares.

Provisional names will change as more specimens are studied and compared. In some cases, what is initially recognized as one species will be divided into two after the full range of variation is observed, and other groups will be lumped. As changes are entered in the computer database, prior names should be stored in a separate field to maintain a history of provisional identifications.

Return Visits

Returning for further collection and further study is crucial during the first two hectares. The taxonomists should expect to visit each quadrat at least three times, maybe more. Return visits are organized using the taxonomic database by, for example, making a print-out of all plants that are missing collections, or all Leguminosae.

One reason for return visits will be simply to collect anew in cases where the initial specimens were insufficient. Also, deciduous or damaged individuals can be checked for new leaves. Another important reason is to study the plants in the field after provisional morphospecies have been assigned. Since the ultimate goal is naming individuals without collecting, this is crucial. Perhaps, for example, those two *Matisia* species that were separated in the lab on the basis of fine pubescence on the petiole also have clear distinctions in branch or trunk form. Return visits can also be used to concentrate on particularly difficult groups.

Lianas

When the taxonomists recognize lianas that were mistakenly tagged by the mapping teams, they should remove the tags. If certain species are frequently confused, a tag with a "V" can replace the regular tag, to ensure that the same plant will not be tagged in future censuses. Of course, tags would not be removed from lianas if it was decided to include them in the census (see chapter 2.2), or if there is some question about which species are truly lianas.

Phase 3—Completing 50 Ha

Once all the individuals in two hectares are sorted into morphospecies and given provisional names, a large part of the identification work is accomplished. Two hectares include 65% of all the species found in 50 ha (Condit et al 1996). The taxonomic crews should make a great effort to learn as many of these 65% as possible before proceeding, focusing on the common species. If they learn to recognize only the most common half of the species, they won't have to collect the vast majority of plants in the remaining 48 ha.

Still, the basic rules for the first two hectares will have to be followed throughout the plot. Unknown plants will have to be collected, and all individuals from confusing groups must be collected. Careful notes must be kept on all. The computer database is expanded as needed by copying new portions of the main database.

When to Stop Collecting

Collecting a particular species is no longer necessary when it is certain that it can be recognized in the field, ideally, when there is no possibility that a yet-to-be-seen species could be confused with it. This decision has to be made according to the judgment of the taxonomists. One general rule to follow would be to never stop collecting until at least 10 specimens from a hard-to-separate morphospecies have been studied. Until a substantial number of individuals from a group have been studied, the range of variation and the key segregating characteristics cannot be known.

Collecting should continue beyond what initially seems needed. It is better to be conservative and collect more than necessary than to stop collecting too soon. It is, of course, always possible to return and collect individuals later, however, if one becomes too certain about field identification too early in the process, one may miss subtly different species that appear later.

Precise lists of which species and which groups need to be collected should be maintained and regularly updated. The chief taxonomist should review these periodically with the field workers.

Communicating

If there is more than one taxonomy team, they should meet to discuss difficult specimens every day, and they should work together while sorting. Lists of which species require collecting should be reviewed jointly.

Species Codes

A system of abbreviations for species is useful. Six-letter codes, comprised of the first four letters of the genus and the first two of the species, are easy to remember. When two genera have the same initial letters, then the fourth position should be a number, with the genus that is alphabetically first getting the lower number. For example, *Trichanthera* and *Trichilia* at BCI have genus codes TRI1 and TRI2. Likewise, two species in the same genus might get one letter and a number.

Once species become well-known, their identity can be recorded solely by their code, without full Latin names. Species codes are used in the database (chapters 3.1-3.3).

Scheduling

Ideally, enough teams work on the taxonomy to keep up with the mapping phase. This has not always proven possible, though, in plots with difficult flora. Well-trained taxonomists are difficult to find, and few students have the skill and interest needed. With fewer teams, the identification phase lags behind the mapping phase. This presents the irritating problem that some mapped plants die before they are identified, nevertheless, these should be kept in the database. As long as the lag between mapping and identifying is kept under a year, only a small number of plants will be lost in this way (annual mortality is around 2% per year; Condit et al 1995a).

Well-Known Flora

My hope with presenting detailed methods for a difficult flora is that it will be obvious what short-cuts can be taken where the flora is better known. If a flora is well-known and not especially diverse, there may be taxonomists available who can confidently name most of the plants in the field. If so, most of the above recommendations are moot. Taxonomy teams can take the data sheets already filled in by mappers into the field and simply fill in species codes. It may even be convenient to have a taxonomist accompany each mapping team, recording species names on the first pass.

Many sites will be intermediate in taxonomic difficulty, with some species or families thoroughly studied and others not. The chief taxonomist must decide which groups can be named in the field and train the teams accordingly. In these situations, the protocol described above for two hectares would be followed, but many plants need not be collected.

Rationale for the Methods and Alternative Approaches

During the first census at both Barro Colorado and Pasoh, an expert taxonomist at each site (R.B. Foster at BCI, K.M. Kochummen at Pasoh) trained the mapping crews to identify the common species (Manokaran et al 1990). Specimens which the mappers did not recognize were flagged and later checked by the experts. We have since decided that this was a mistake, even at BCI, which has a relatively well-known and not especially diverse flora (Croat 1978). Most of the mappers are students in botany or zoology, or young forest technicians who only know timber groups. Although the common plant species can be learned quickly, there are rare species which any given field worker may seldom see (half the species comprise fewer than 2% of all stems; Hubbell in press; Condit et al 1996). The problem is that a rare species, never before seen by an inexperienced worker, might be mistakenly labeled as a familiar species. In addition, forest technicians sometimes know various local names for a given species, and two mappers sometimes use different names for the same species. These problems cannot be solved without the taxonomic leaders checking the specimens.

Because of the error rate in 1982 at BCI, it was decided in 1985 and 1990 that R. Foster would identify every single new plant—35,000 in each census. By 1995, four Panamanian botanists, each with more than five years experience at BCI, took over the chore and were responsible for identifying the new plants. Thus, a group of five experts has checked the identities of all new plants in the plot (after 1982), and they have also rechecked the original identifications on certain difficult genera and families. I currently estimate that <1% of the individuals in the plot are misidentified (see chapter 2.4 on error-checking).

Plant identification at most of the large plots is now being done entirely by experts, but the exact procedures have varied from plot to plot. At Lambir, J. LaFrankie had mappers collect sterile specimens from all individuals, and LaFrankie and P. Ashton sorted and named these at the field lab. S. Tan visited all trees ≥ 100 mm dbh in the field. The procedure for full collecting in two hectares described above is exactly what R.B. Foster has instituted at Yasuní in Ecuador; two taxonomists-in-training are now working on their own.

At the Ituri plot in the Congo and the Sinharaja plot in Sri Lanka, taxonomic work was done along with mapping work. Large field teams included taxonomists who had trained with more experienced botanists in the area. Still, much collecting and revising was necessary, as described above.

The taxonomy work could keep up with the mapping work in the Congo and Sri Lanka largely because the taxonomists had done considerable collecting before the plot started. Indeed, Manokaran et al (1990) recommended that 75% of the local tree flora be documented before the plot begins. Then identification in the plot can proceed quickly, and relatively little collection is necessary. There are definite advantages to this approach, but if time and funds are limited, it may not be reasonable to delay the mapping work while extensive collecting is carried out. The first two hectares of the large plot become the training ground.

Time and Labor

It is difficult to make exact estimates of the time and labor needed to complete the identification phase, because much of the work has been carried out by senior taxonomists who work irregularly while in the midst of dozens of other projects. It also depends a great deal on how well the flora of a site is already known.

At Ecuador, I estimate that the taxonomy—which is being done completely separately from the mapping—will require about three years for seven people, including part-time work from R.B. Foster. This comes to 5300 person-days. Combining this with 5000-7000 days to finish the mapping (chapter 2.2), and assuming 21 work-days per month, 500-600 person-months will be necessary to finish the 50 ha. Relative to other plots, this is heavily slanted toward the identification work, because so little prior collecting work had been done in the area.

In both the Congo and Sri Lanka I can only estimate the labor needed for mapping and taxonomy together, since teams worked on both simultaneously. In the Congo, 298 person-months were needed to fully map and identify the 40 ha plot. In Sri Lanka, 320 person-months were needed to finish the 25 ha plot. Neither estimate includes the topographic work (see chapter 2.1).

Key issues for the taxonomy

- Experts and keen botanical students trained by experts do all the identification
- Thorough collecting must be done if the flora is poorly known, and a field herbarium is necessary
- Two hectares should be completely identified before proceeding; nearly every specimen in the two hectares is collected
- Taxonomists maintain their own computer database of field notes and provisional names for all specimens; it is based on a copy of the main database
- Collecting should continue until a wide range of specimens of all difficult groups are studied
- Taxonomists must be careful not to become too complacent in their knowledge—the worst error is to miss subtly different species
- The eventual goal is naming most specimens without collecting.

CHAPTER 2.4

Checking the Field Work

With such a large census, and teams of mappers who are often inexperienced students or technicians, strict protocols for checking and revising all aspects of the data collection are important. Supervisors must carefully check the work of the field crews regularly, particularly near the beginning of the census. As long as one supervisor is always working with each field team, this is accomplished as part of the daily mapping routine and requires no further discussion. More important, there should be systematic re-sampling of stratified random samples of trees, aimed at both revealing errors so that improvements can be made and at estimating error rates for reporting purposes.

Phase 1—Dbh Checks

It is easy to perform blind checks of dbh measurements simply by re-measuring randomly selected plants. Each field worker is assigned a quadrat measured < 3 months earlier by a different mapping team. Working alone, the second worker records dbhs a second time, without the benefit of knowing what the first worker recorded. In addition, the quadrats are selected at random, so the teams never know which of their quadrats will later be re-measured (so it is a double-blind). All trees ≥ 100 mm dbh in the quadrat and all trees ≥ 10 mm in the first two 5 x 5 m subquadrats (1,1 and 1,2) are re-measured. At BCI, this approach includes, on average, 22 trees < 100 mm dbh and 17 trees ≥ 100 mm dbh per quadrat, thus providing a balanced sample of large and small stems. With 10 mappers each re-measuring two quadrats, about 400 plants in each dbh group are measured. Figure 2.4.1 shows the results of a dbh check carried out at BCI.

There are two ways in which this quantitative revision should be used. First, it can be done at the very start of a census to determine whether improvements are necessary. One gauge of accuracy would be to compare to our BCI results, where the mean deviation for plants < 100 mm dbh was 4.2%, and for plants ≥ 100 mm, 0.27%. Another guide for accuracy is to consider growth rates: abundant, shade-tolerant trees at BCI grow 6-18% per year when saplings, and 1-4% per year as large trees; pioneer species grow even faster (Condit et al 1993a). I suggest that the accuracy of dbh measures should be less than these figures, so that after five years, the mean increments exceed the degree of accuracy by 5-fold. Thus, the goal should be < 6% error in saplings and < 1% error in larger trees. If an estimate of accuracy early in the census suggests that these levels are not being reached, then the mappers should revise their techniques.

Tropical Forest Census Plots: Methods and Results from Barro Colorado Island, Panama and a Comparison with Other Plots, by Richard Condit.

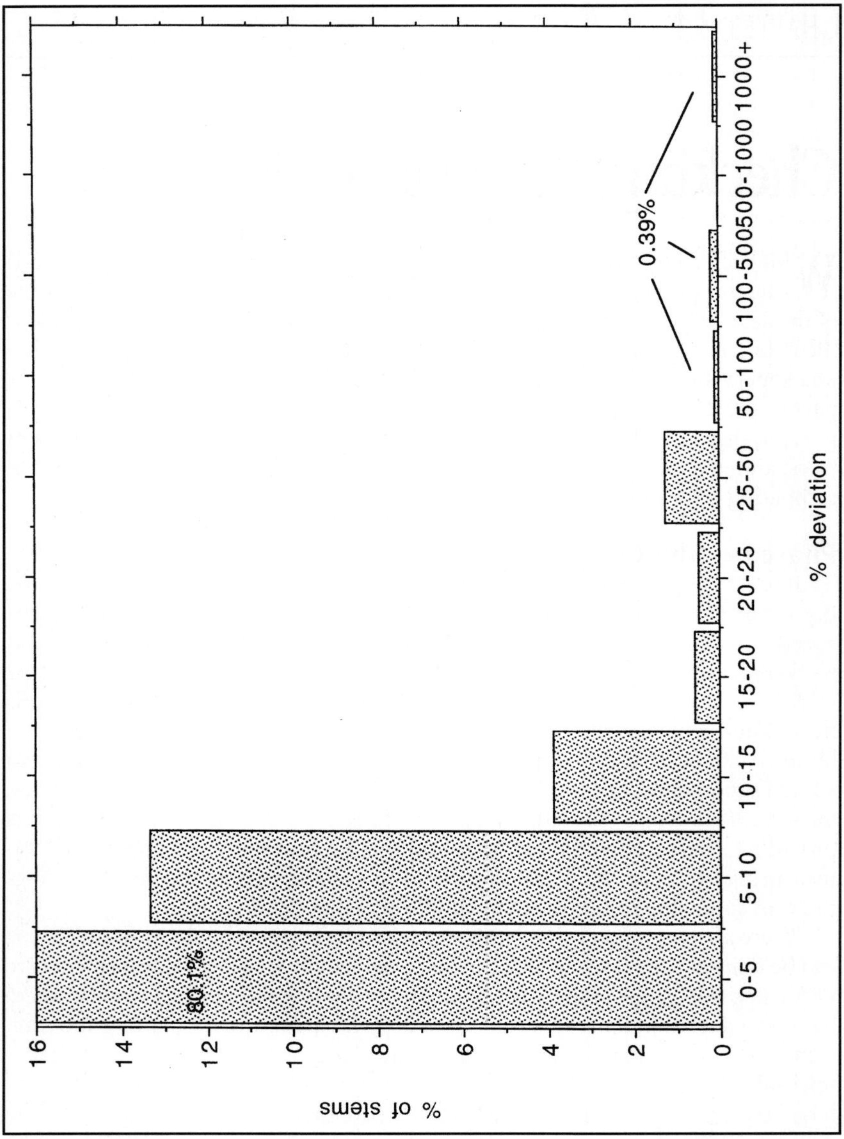

Fig. 2.4.1. Error rate in dbh measures among 1033 randomly checked plants in the Barro Colorado 50 ha plot during the 1995 census. A total of 0.39% (4 plants) had discrepancies ≥ 50% between two independent measures.

The error-check procedures should be repeated during the second half of the census when most work is complete, simply as a record of accuracy. At this stage, the census-workers are experienced and have presumably reached a plateau of accuracy; it is too late to correct their techniques anyway.

Phase 2—Taxonomy Checks

Error rates for plant identification should also be estimated in randomly selected quadrats or in random samples of individuals. This is best done when the entire census is finished and all the data, including species identifications, entered in a database. At BCI, we checked all 18,694 plants in 84 randomly chosen 20x20 m quadrats and found 159 wrongly identified (0.85%). An alternative procedure for checking taxonomy is to focus on difficult groups. At BCI in 1996, we checked all individuals in the two species of *Spondias*, and found 7% were misidentified, always as the other congener. In *Protium*, we fully checked 10 hectares, and found that 1.6% were incorrectly called another species in the family; one was misidentified in the wrong family (0.1%). Since these two genera were selected as the most problematic, these error rates are consistent with an overall rate < 1%.

A better approach is to sample individual species. From each species in the entire plot, 4-5 individuals are re-identified (fewer for species with < 4 individuals in the plot). The error rate can then be tallied by abundance category. This allows an error rate for rare species to be checked. In contrast, sampling entire quadrats gives an error rate mainly for the common species, and rare species may pose more problems. This proved to be the case at BCI: of 400 individuals of the 100 most common species, 0.75% were wrongly identified, while of 706 individuals of the 200 least common, 2.41% were wrong.

Alternative Approaches to Checking

At Yasuní in Ecuador, the mapping crews began the census working individually, not in teams. In this circumstance, we wanted the supervisors to visit each and every quadrat to survey the work (as we do during recensuses at BCI; see chapter 2.5). At the start of the census, the supervisors re-measured many dbhs and immediately reported errors to the mapper who had worked the quadrat. Later, they were less thorough, but still visited each quadrat, screened for missing plants, and checked difficult stems. This is one reason I now suggest that solitary mapping work is less efficient than teams—with teams led by a supervisor, these return trips are obviated.

Many plots have carried out random checks for dbh errors and identification errors. At Pasoh, Manokaran et al (1990) reported similar error rates to what we found at BCI, and found that painting the stems reduced the error. Similar error rates were also found in the Congo plot.

Taxonomy is easier to check, since it can be done at any time (since plants do not change species). At Luquillo, every plant was re-identified during the second census, and the field workers were not told the initial identification. This provided a blind estimate of the error rate, but the results are not yet tallied. Error rates at the two Malaysian plots were quite a bit higher than what we found at BCI, but these plots are much more diverse than BCI, with many very difficult species.

Time and Labor

These checking procedures require very little time. Each mapper should spend two days re-measuring dbh, once early in the census and then again near the end. With 10 mappers, this requires 40 person-days. The taxonomists can re-identify a sample of 1000 plants with about 20 days in the field. The total labor expended is thus about three person-months.

Key issues when checking the census

- Mappers should get feedback on errors very early in census
- Stratified samples should be checked to estimate error rates for dbh and species identity

CHAPTER 2.5

Recensus

The overall goals of the recensus are to revisit all stems tagged and mapped during the first inventory, record their status, and to tag and map all new recruits—those plants which grew past the 10-mm size limit. Most of the methods are identical to those from the initial census.

There are a couple of new issues which require careful attention during the recensus, though. One is how to relocate the exact point where trees were measured, and the second is how to establish whether trees are dead. Both require precise rules that must be followed rigorously. There are other minor adjustments to the methodology, and I also cover in detail various errors that are unique to the recensus. I follow the order of presentation used in chapter 2.2, but I also cover the checking and taxonomic work in this chapter.

Census Interval

Because the goal of the CTFS network of large plots is to create comparable data sets based on standardized methods, most existing plots plan to utilize the same five-year census interval. Although five years is short in terms of the trees' lifetimes, there are a couple of considerations favoring brevity. First, the longer the interval, the more short-term fluctuations in demography will be missed. Second, longer intervals will lead to more lost tags and thus more lost trees, and whereas it is easy to relocate large trees after many years, this is not so true for small saplings. On the other hand, big plots are costly to recensus, and the five-year interval was selected as a compromise between the biological need for frequent censuses and the practical difficulties associated with organizing a work force and finding funding.

The first census ordinarily takes up to three years, and the five-year interval should apply to the census mid-points. Thus the second census might begin only three years after the first one ended.

Before the Recensus

It is essential not to forget the plot during the five-year interval. Some stakes at the 5- and 20-m grid points will be buried in fallen vegetation or knocked over. The 20-m stakes should have been pounded in so deeply that a falling tree might break them but not remove them: thus, the location is maintained, but the stake should be replaced. If the stake is moved, it must be carefully replaced using a

Tropical Forest Census Plots: Methods and Results from Barro Colorado Island, Panama and a Comparison with Other Plots, by Richard Condit.

surveying compass. Bright flagging used to mark 20-m stake positions will certainly need replacing after two years. The 5-m stakes can be replaced using a tape measure, and mappers can do this as they arrive at each quadrat.

It is important to check the stakes about two years after the first census is over, since it is our experience at BCI that many stakes will be damaged or displaced by then. If one waits the full five years, then the effort and the errors will be magnified. Replacing single missing stakes is easy, but when several neighboring stakes are lost, more work is required.

Personnel and Work Teams

Most recommendations from the initial census apply. Apart from the chief scientists, there are field supervisors and mappers; however, the mappers should work alone during a recensus because far less tagging and mapping is needed. The field supervisors will be responsible for checking every quadrat after the mappers have finished. Since far fewer plants need identification, and the flora is well known, the taxonomy work should be done with the checking: the supervisors do both.

Equipment

Field workers need the same equipment for the recensus as they did for the first census: calipers, diameter tape, a compass, clipboard, paper, and pencil. Binoculars, a ladder, and large calipers are needed by the big-tree mapping team.

Plants to Include

All rules given in chapter 2.2 are followed in the recensus. There is one exception which will be addressed more thoroughly below: plants from the first census whose stem ≥ 10 mm dbh broke, and which have no current stem ≥ 10 mm dbh, are maintained in the census.

Data Collection

Data Sheets

The five types of data sheets used in the initial census are all used again in the recensus (Fig. 2.2.1). Three of these sheets are unchanged. The main data sheet from the first census (Fig. 2.2.1B) remains, although it is now called the "recruit" data sheet, because only data from those newly appearing plants—the recruits—are recorded on it. Whereas three or more main sheets were needed per quadrat during the first census, only one recruit sheet is needed during the recensus, because there are usually fewer than 30 recruits per quadrat after five years. Multiple-stem and problem data sheets are also identical to those used in the first census (Fig. 2.2.1).

The maps and big-tree data sheets are changed slightly. Maps are not blank, but have plants from the prior census included. In addition, maps of entire 20 x 20 m quadrats can be used, since relatively few plants need to be added (Fig. 2.5.1). I recommend using maps generated by computer from the first census database, indicating each tree by a circle with the tag number adjacent (only the last three digits are necessary; see chapter 3.2). The big-tree sheet also includes prior measurements of big trees from the first census: both the diameter and the height at which they were measured. Otherwise, it is just like the sheet from the first census (Fig. 2.2.1).

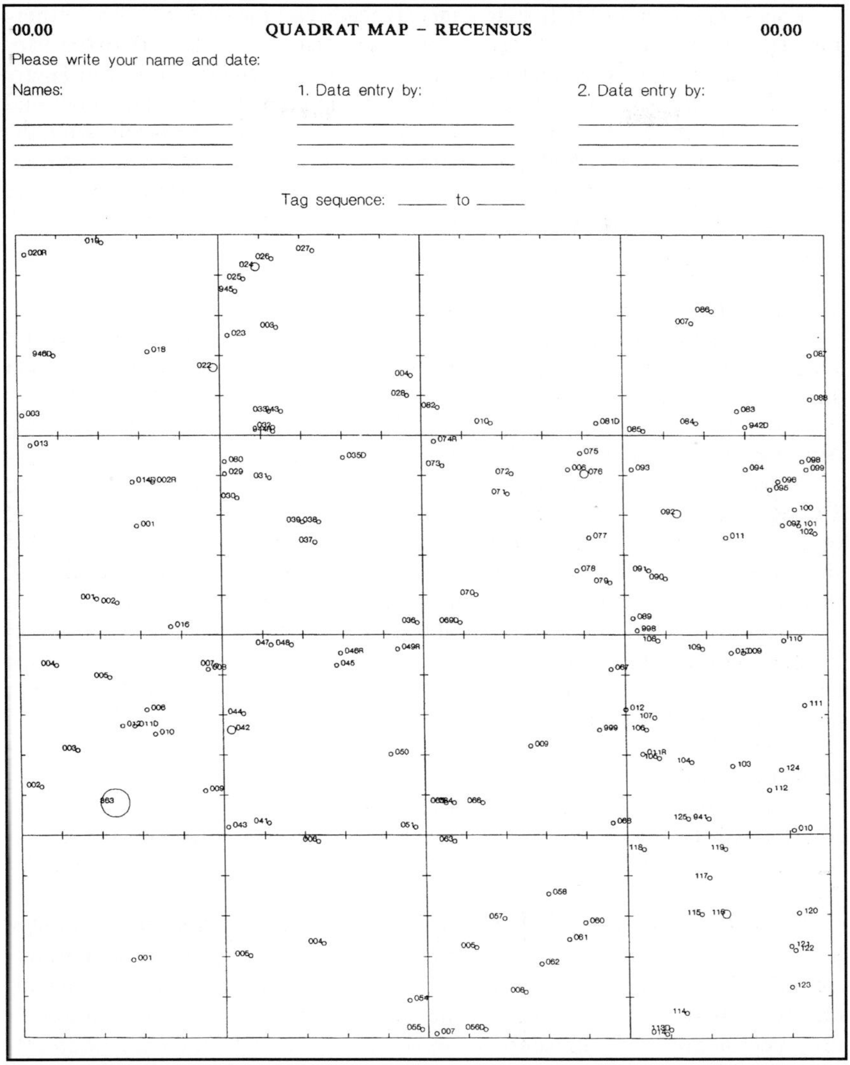

Fig. 2.5.1. Field map for recording the location of new recruits during a recensus. This is similar to the field map used during the first census (Fig. 2.2.1), but it has locations of previously mapped trees entered by the computer, and the full 20 x 20 m quadrat is on one sheet.

Finally, one entirely different data sheet is needed for the recensus. This is where new data are recorded on the original plants, those tagged and mapped in the prior census. This is called the "original" data sheet, and a sample is shown in Figure 2.5.2. This data sheet is like the recruit data sheet, but it has the tag number and species code already filled in for each plant, along with the dbh and codes from the prior census. For plants whose point-of-measure (POM) was not breast height, the height-of-measure appears on the sheet with the prior codes. All this information is important to the mappers during the recensus. The new dbh and codes are recorded in the blanks.

Work Sequence

As in the first census, mappers (now working alone) finish one column at a time, starting at the south end. A clockwise pattern should be followed within each subquadrat (Fig. 2.2.2).

Mapping the Trees

Procedures for mapping trees are identical to those used in the first census. The big difference is that only the recruits need be mapped, and the position of original trees on the map can be used as a guide. All new trees which are mapped should be marked with an orange flag, so that the supervisors can quickly find them when they check the quadrat and make the identifications.

Mappers should not remap original trees, even if they think the original map position is in error. If a mapper believes a tree was mismapped by more than 1 m, it should be marked as a problem for the field supervisors to check (and thus marked on the problem data sheet).

Tagging Trees

Procedures for tagging in the recensus are identical to those from the first census. However, there is a very important caution: before tagging a new plant, it must be confirmed that it does not have an old tag. Mappers must be sure not to double-tag!

Since only about 10% of the trees need tags after five years, fewer numbers need be assigned to each column. With columns 500 m long, as in 50 ha plots, it is convenient to assign 1000 tags per column, just as it was convenient to assign 10,000 tags per column in the first census. If the first census used tags up to 449000, then the recensus can start with 500000-500999 in column 00, 501000-501999 in column 01, etc. Some columns will have more than 1000 new trees, in which case tags will have to be borrowed from a prior column that had fewer than 1000. This means that a few tag numbers will not correspond with column numbers.

Mappers should also replace tags lost from original trees. Sometimes, the string breaks when the tag is lifted, and extra pieces of string need to be carried for this circumstance. Tags attached with nails should be replaced if the nail head is engulfed by the trunk or is close to the bark. In these cases, string rather than a nail should be used for the new tag, since the tree must be growing rapidly.

In some cases, tags will be missing altogether. If a mapper suspects that a tree is missing a tag, it should be indicated on the problem sheet and left for the supervisors. For trees ≥ 100 mm dbh without tags, the supervisor can easily confirm that the plant had been tagged and find the correct number by consulting the map. He or she then must engrave a new tag with the original number and return later to tie

40,02 ORIGINAL PLANTS - RECENSUS 40,02 (1)

Names: Please write your name and date:
1. Data entry by: 2. Data entry by:

Tag	Sp	Dbh		Tag	Sp	Dbh		Tag	Sp	Dbh	
1,1											
001572	CECI	456B200	___ ___	055271	HYBP	21	___ ___	055278	HYBP	16	___ ___
055265	CHRP	92	___ ___	055272	HYBP	52	___ ___	055331	HYBP	19	___ ___
055266	HYBP	31	___ ___	055273	PYHO	23	___ ___	305038	HYBP	10	___ ___
055267	SWA2	76	___ ___	055274	PYHO	35	___ ___	305039	PYHO	12R	___ ___
055268	PYHO	10R	___ ___	055275	HYBP	19	___ ___	305040	PYHO	18	___ ___
055269	HYBP	17	___ ___	055276	LACA	86	___ ___	440042	ALSB	14	___ ___
055270	LACA	82	___ ___	055277	PYHO	28	___ ___				
1,2											
055279	SWA2	17	___ ___	055282	FARO	64	___ ___	305041	PYHO	15	___ ___
055280	HYBP	21	___ ___	055284	HIRT	20	___ ___				
055281	HYBP	19	___ ___	055285	OENM	95	___ ___				
1,3											
055286	HYBP	48	___ ___	055291	PROC	66	___ ___	055296	SYMG	-1D	___ ___
055287	ALSB	43	___ ___	055292	DESP	59	___ ___	305045	DESP	14	___ ___
055288	PTER	26	___ ___	055293	HYBP	18	___ ___	440043	DESP	15	___ ___
055289	HYBP	34	___ ___	055294	TRI3	-1R	___ ___				
055290	DESP	26	___ ___	055295	ALSB	62	___ ___				
1,4											
055298	HIRT	107	___ ___	055305	FARO	45	___ ___	440045	XYLM	10	___ ___
055300	DESP	22	___ ___	055306	XYLM	30	___ ___	440046	HYBP	11	___ ___
055301	HYBP	18	___ ___	055307	HYBP	-1D	___ ___	440047	HYBP	12	___ ___
055302	HYBP	38	___ ___	055308	ASPC	41	___ ___	440048	ALSB	14	___ ___
055303	GUEF	14	___ ___	305042	TRI3	16	___ ___	440049	HYBP	12	___ ___
055304	CAPF	14	___ ___	440044	CHRP	12	___ ___	440050	HYBP	10	___ ___
2,1											
055329	RHEE	74	___ ___	055335	CUPS	-1R	___ ___	305049	ALSB	14	___ ___
055330	UNOP	111	___ ___	055336	ALSB	24	___ ___	440058	HYBP	18	___ ___
055332	HYBP	24	___ ___	305046	HYBP	23	___ ___	440059	PROT	11	___ ___
055333	ALSB	89	___ ___	305047	GUA2	34	___ ___				
055334	PYHO	51	___ ___	305048	HYBP	13	___ ___				
2,2											
001576	ZANB	382	___ ___	055325	HYBP	-1D	___ ___	055327	ALSB	37	___ ___
055324	OENM	107	___ ___	055326	HYBP	46	___ ___	055328	HYBP	30	___ ___
2,3											
055297	PTER	20	___ ___	055321	BROA	58	___ ___	440054	HYBP	12	___ ___
055318	HYBP	43	___ ___	055322	ALSB	23	___ ___	440055	ALSB	15	___ ___
055319	GUA2	61	___ ___	055323	XYLM	28	___ ___	440056	DESP	14	___ ___
055320	POUU	17	___ ___	305044	HYBP	11	___ ___	440057	ALSB	10	___ ___
2,4											
055309	CAS1	51	___ ___	055314	DESP	34	___ ___	440051	HYBP	11	___ ___
055310	HYBP	24	___ ___	055315	OLMA	95	___ ___	440052	TRI3	13	___ ___
055311	HYBP	19	___ ___	055316	XYLM	31	___ ___	440053	HYBP	10	___ ___
055312	TROR	-1D	___ ___	055317	AEGP	103	___ ___				
055313	HYBP	24	___ ___	305043	TRI3	-1D	___ ___				

STRI.BCI.94.1v, 15DEC94.

Fig. 2.5.2. Field sheet for recording data on the original trees—those already tagged and mapped in a prior census.

it to the tree. If small saplings do not have tags, supervisors must consider the situation carefully: if the sapling matches the species, location, and dbh of a mapped plant, and there is no sign elsewhere of this mapped plant, then it is reasonable to assume that the tag was lost. (In my experience at BCI, very careful searching through the litter will produce the missing tag in this situation more often than not.)

Defining and Establishing Death

A tree is recorded as dead if it has no living leaves or sprouts at all and is clearly dead, or if it is completely missing (whether or not the tag is found in the litter). Dead trees are not measured.

A tree is recorded as alive if there are any signs of life—no matter how small a sprout from no matter how large a trunk. Deciduous trees which have lost all their leaves should be recorded as alive, and mappers should be taught to look for deciduousness and should investigate leafless individuals closely for living sprouts and leaves, or for signs that the wood is rotting. It seems inevitable that a few living trees will be called dead though, and during later censuses, they can "come back to life": at BCI, 123 trees that had been called dead in 1990 were found alive in 1995, of 26,933 plants recorded as dead in 1990 (Condit et al 1995a).

When dead trunks are found, or at least their tags, the mapper's job is easy. When there is no sign whatsoever of a tree nor its tag at its mapped location, the mapper must carefully search. Mappers must be aware that some plants were mismapped in the first census, so each subquadrat must be thoroughly searched, and adjacent subquadrats should also be checked if the plant is near a boundary.

Measuring Trees

For the most part, measuring trees during a recensus is just like measuring trees during a first census; indeed, the whole point is to use the same procedure so that data match. New recruits are measured following the rules outlined in chapter 2.2. Original plants must be remeasured at the same place they were measured before (thus, the POM or point-of-measure must match). Problems with changing POMs are the most difficult part of the recensus, and I add new sections here to address them.

1. Finding the POM

For every tree recensused, a mapper first consults the original data sheet to see whether there is a code 'A', meaning the plant had a POM different from breast height, or 'I', meaning that no good POM was found. In either case, the tree should have been paint marked, and the height of the POM should appear on the data sheet with the original codes (Table 2.2.1). The mapper must then decide whether the prior POM—whether at breast height or a different height—is still usable. If so, then there is no problem, and a measurement is taken using the rules of the first census. The code 'I' for irregular stem or 'A' for alternate POM must be re-applied when appropriate, even if the plant had 'I' or 'A' in the previous census.

There are two ways by which the prior POM might no longer be usable: 1) the main stem broke but sprouted, or 2) a new swelling or buttress reached the POM. In each case, a new POM must be chosen, and the code 'C' must be recorded to indicate a change of POM. If necessary, new codes 'A' or 'I' must be applied, and the height of the POM recorded.

2. Resprouts

If a plant broke its main stem since the prior census, but did not die, it is called a 'resprout'. The code 'R' is applied, plus the code 'C' indicating a change of POM. The new stem is measured as usual at breast height. If there is no stem with a dbh ≥ 10 mm, the plant is given a dbh of 0. It remains part of the census and keeps its tag.

Unfortunately, resprouts cause confusion, principally because it is not always possible to tell whether a stem broke since the last census or prior to the last census. This is why the code 'X' was used in the first census to indicate a stem broken below breast height (Table 2.2.1). When the mapper sees this code from the first census, he or she knows not to add the code 'R' (unless it appears that a second break occurred since). If there was an original code 'X' and no evidence for a new break, 'X' should be re-applied.

There can still be questionable cases, for example, if there is an obscure indication of a break, like a lump on the side of the stem, but no code 'X'. In equivocal cases, the prior dbh can be consulted, but only with caution. Thus, if the living stem has a dbh much smaller than the original dbh, the resprout code is indicated. Additional guidance can be taken from the size of the living stem (if it is larger than 30 or 40 mm, it is probably not a new sprout) and from the size of the broken stem, which should match the original dbh. However, the resprout code should not be applied simply because the current dbh is smaller than the original dbh. It should not be applied indiscriminately to cover up errors in measurement, but only if there is additional evidence for stem breakage.

This is a very important point which recensus mappers must understand well. If a new dbh is smaller than an old dbh, it still *must* be recorded. We have to accept errors when they are there. Statistical analyses will handle them.

3. New Swellings and Buttresses

If a stem was previously measured at breast height, but now has a buttress or swelling there, an alternate POM must be chosen according to the rules outlined in chapter 2.2. The codes 'A' and 'C' must also be added. If the newly ascended buttress is too high, then the tree joins the rank of big trees and is measured later by the big-tree team.

4. Big Trees

Big trees with buttresses at breast height are simply marked 'B' during the main census, and a big-tree team with a ladder measures them later, just as in the first census. These trees were supposed to have been measured 50 cm above the highest buttress (chapter 2.2), and the team has the big-tree data sheet giving this original height. If the original POM is still ≥ 30 cm above the highest buttress, then a new measurement is taken at the same spot. The code 'C' is not applied, since the POM has not changed.

However, if the original POM is < 30 cm above the buttress (because the buttress ascended more than 20 cm), then two diameters should be taken—one at the original POM and one 50 cm above the buttress. The height of each must be recorded. The code 'C' is not applied, because the original POM was used. The higher diameter is stored away for future reference, in case the buttress ascends even higher by a later census.

Multiple-Stemmed Individuals

Multiple stems are handled exactly as in the first census. The code 'M' goes on the original or recruit data sheet in the code column. For recruits, the tag must be placed on the largest stem. All dbhs are recorded on the multiple-stem sheet.

The largest stem was indicated during the first census by placing the tag around it (or perhaps with paint). This is the key stem: it must be measured at the same POM as previously, and if it cannot be, for any of the reasons described above, 'C' is applied for the entire plant (on the main data sheet). The resprout code would also be applied if the largest stem broke and grew a new sprout. Changes in POM on secondary stems are ignored, although each is measured.

In some cases, a secondary stem may overtake the primary one and become the largest stem on the plant. The tag is switched to the largest stem and the code 'C' applied to indicate the change of POM.

Palms

Rules for palms are no different from the first census, and palms pose no unusual problems.

Stranglers

Rules for measuring stranglers are also identical to the first census. Since stranglers have irregular trunks, care must be taken to match unusual POMs.

Summary

Describing what data go on the data sheets provides a summary for everything recorded during a recensus. As in the first census, there is one complete set of data sheets for each 20x20 m quadrat.

For recruits, data are recorded exactly as in the first census. They are indicated on the map with a circle and the last three digits of their tag, and the range of new tags used in a quadrat is written across the top of the map (Fig. 2.5.1). On the recruit data sheet, tag number, subquadrat number, dbh, and codes are recorded. Only the codes used in the first census apply (Table 2.5.1). When a recruit has multiple stems, 'M' is placed in the code column, and if it requires a ladder for measuring, 'B'. As in the first census, mappers should use the problem code for situations they do not know how to handle, or for problems which they can handle but which require later attention (e.g., a lost tag).

In addition, as in the first census, mappers should be encouraged to fill in their best guess for the identity of each recruit. This helps them learn, and allows the identification team to assess how well mappers know the flora. It is especially helpful during a recensus for mappers to learn some species, since it can help a great deal in re-locating stems from earlier censuses. Mappers might then also catch identification errors from earlier censuses (the mapper would mark a misidentified tree on the problem sheet so that the taxonomists would check it).

The original data sheet already has a tag number, subquadrat number, species name, and codes from the prior census (Fig. 2.5.2). A new dbh and new codes are entered. These plants are not placed on the map, since they already appear there. Any of the codes listed in the prior paragraph can be applied, as well as one or more of the recensus codes—'D' for dead, 'R' for resprout, and 'C' for change-of-POM (Table 2.5.1). Dead plants do not get a dbh. Multiple-stem plants are given 'M' in the code column, and big trees a 'B'. The problem code should also be used when needed.

The problem data sheet is used as in the first census, to record comments about each problem. Each plant receiving a 'P' code on the recruit or original data sheet must appear on this sheet. Some problems will be resolved by the mapper on a subsequent day. Others will go to the supervisors. Problems are crossed off the problem sheet when resolved, and the 'P' code crossed off the main sheet.

Multiple-stem data sheets are used just as in the initial census. They are completely blank to begin with, and new dbhs of all 'M' plants, whether original or recruits, are recorded. Every plant getting a code 'M' on either the original or the recruit data sheet must have dbhs recorded on the multiple-stem sheet.

Table 2.5.1. Codes used during the recensus of a large plot. These are entered on the data sheet for recruits and original trees. The first seven, through code 'X', are also used in the initial census (Table 2.2.1). Codes C, R, and the death codes can only be used in a recensus.

code	use	where recorded	additional information
B	large buttress, requiring ladder to measure	code column	dbh entered on big-tree sheet
M	multiple stems	code column	dbh entered on multiple-stem sheet
A	POM at alternative height, not breast height	code column	height of POM entered with code column as well
I	stem irregular where measured	code column	
P	any problem requiring further attention	problem column	description on problem sheet
L	stem leaning	code column	
Q	stem broken above breast height	code column	
X	stem broken below breast height	code column	
C	POM has changed since prior census	code column	usually used with R, A, or I
Y	prostrate stem	code column	height of POM entered with code
R	resprout (main stem broken but resprouted since last census)	code column	dbh 0 is permitted if new stem < 10 mm
DS	dead, stem standing	code column	no dbh entered
DC	dead, stem fallen	code column	no dbh entered
DT	dead, only tag found	code column	no dbh entered
DN	presumed dead, no tag nor stem	code column	no dbh entered

The big-tree data sheets are the same as those from the first census, but they include the original diameter and height-of-measure for every big tree from the first census. New dbhs and heights are added, and while in the first census only one diameter was entered, big trees sometimes get two diameters in a recensus (thus the two diameter columns on the big-tree data sheets; Fig. 2.2.1E). There may be new big trees as well, cases where a ladder was not necessary in the first census but was during the recensus, and these are recorded on the same sheet. Every tree which got a 'B' on the original data sheet must appear on the big tree data sheet.

Taxonomy and Checking

As for a first census, I recommend that all identification be done by experienced botanists (see chapter 2.3). Unlike the first census, I recommend that the plants be identified at the same time as the quadrats are checked by supervisors.

Thus, the taxonomy teams and supervisors are one and the same. Ideally, the supervisors are the same people who worked on the taxonomy during the first census, otherwise, they will need new training in identification.

The supervisors work alone, visiting each quadrat a few weeks after the mapper has finished. This contrasts with the first census, when supervisors worked along with the mapping teams. They re-measure some dbhs, check multiple stems, look for recruits that the mapper missed, and resolve problem codes. At the same time, they find every new recruit—each marked with an orange flag by the mapper—and enter its species name on the recruit data sheet. The orange flag is removed when the plant is identified.

There is a new twist to the problem list during the recensus which the supervisors must handle: some of the problems will involve errors from the original census. Most frequently, these are mismapped or misidentified trees. If a supervisor believes a plant was mismapped by 1 m or more, a red arrow should be recorded on the map indicating the shift, and the plant's tag and subquadrat entered on the problem data sheet, along with a comment on the situation. Any plant misidentified in the first census must also be recorded on the problem sheet, with a new species name included. In addition, tagging errors from the first census, such as two plants given the same number, should be corrected in the field, and an account must be recorded on the problem sheet.

These errors remain on the problem sheets and are later given to the data entry team. They must be corrected in the original database. This contrasts with the first census, when all problems were resolved by the supervisors before submitting the data sheets.

A stratified sample of stems should also be re-measured using the double-blind approach recommended for the first census (chapter 2.4). In addition, a second technique for catching dbh errors can be applied during a recensus, based on growth rates, but only after data have been entered in the computer. After a portion of the census—perhaps the first 25 ha—is in the computer, plants are divided into dbh classes based on the second census: 10-19 mm, 20-39 mm, 40-79 mm, 80-159 mm, 160-319 mm, and 320 mm and above. Growth rates for every plant in the entire plot are then calculated, and these are ranked from the highest to the lowest within each dbh class. The highest and lowest growth rates are usually obvious errors; the lowest growth rates are negative.

The 50 highest and 50 lowest growth rates from each dbh class—600 plants all told—should be re-measured in the field. Because the data are extracted from the computer, all 600 can be sorted by quadrat, and each mapper given a column. Based on BCI experience, as many as half of the 600 extreme growth records will prove to be large errors in dbh, often 10-fold mistakes (like 200 instead of 20). Some of the errors will be from the first census, though, and cannot be fixed. In those cases where the current dbh was obviously mismeasured, however, a new dbh should be recorded. These corrected dbhs will be entered in the computer, and the census date for those plants must be changed.

A Third Census and Beyond

The rules for a third census are no different from those for a second, however, some novel situations appear. Plants reported as resprouts during the second census, even those with dbh 0, must be censused again, so they appear on the census maps, and the code 'R' and the prior dbh must appear on the original data sheets. If plants that got an 'R' in the second census did not suffer another break (this should be evident from stem form), the code 'R' is *not* reapplied (nor the code 'C'). This is a very important rule which must be rigorously taught and enforced!

A third census is also the first time that trees can come back to life. Dead trees from the second census should be placed on the recensus maps—with a 'D' next to their tag number—just for this reason. Mappers in the third census should not search for these trees, but if one of them turns up alive, its tag number and current dbh should be recorded on the problem sheet. By having dead trees on the map, the field workers are given enough information to fully understand these situations.

In a fourth census and beyond, no new situations can arise. Dead plants should be included on maps the first census after they die, but no more. Thus, plants found dead in the second census will appear on maps for the third census, but not the fourth, etc.

Rationale for the Methods and Alternative Approaches

The detailed rules I listed above are designed with two main goals in mind: remeasuring trees at the same POM and accurately censusing dead trees. I fully examine here the rationale for these rules and consider alternative approaches, particularly the methods that have been followed at the BCI recensuses.

Census Interval

A standard interval is important for comparing plots, because growth, mortality, and recruitment rates are affected by the length of the interval (Sheil 1995b). In fact, most of the plots have settled on the five-year period. However, the Mudumalai plot is fully censused every four years and partially censused every year (dead plants and recruits are recorded, but no dbh measurements made). The very high mortality rates at Mudumalai (R. Sukumar, unpublished data) favor a shorter interval. BCI data suggest that the bias caused by differences in census interval is slight, though, at least up to 13 years: mortality over the full 1982-1995 interval was 2.26% per year for trees ≥ 100 mm dbh, nearly identical to the 2.21% found by averaging the three shorter intervals (1982-1985, 1985-1990, and 1990-1995).

At both the BCI and Pasoh plots, the second census was finished three years after the first. This was partly because the first census took such a long time, and we wanted to get in step with censuses every half-decade (i.e., 1990, 1995, etc.), and partly because the second census was seen as a chance to clean up errors from the first (e.g., missing plants, misidentifications). I do not see any need for this in other plots, and I suggest that the midpoint of the second census be exactly five years after the midpoint of the first.

It was initially hoped that all plots would be censused in the same years, every half-decade. This would raise the network's power for detecting global events that affect all the plots. However, it seems likely that practical considerations will intercede, for instance, the war at the Ituri site in Congo. Since important global events are likely to have long-term effects, identical census years is not a high priority.

Mapping

As an alternative to using a computer-drawn map as a data sheet, a Xerox copy of the map created in the field during the first census could be used. The advantage to this is that hand-written comments and features marked on the map would be available to the second mapper, even if they had not gone into the computer database.

The disadvantage is that field maps are generally harder to read than computer maps, and they may include errors that were later corrected. Most importantly, this approach would only work for one census—new maps would surely have to be produced for the third census.

Failing to Locate Living Plants

Imagine yourself searching subquadrat 4,4, the last in a quadrat you have been working in for more than a day. A 10-mm tree with tag 467 does not appear where the map indicates it should. Is it dead? It's late in the day and your cursory check of the subquadrat turns up nothing, so you enter the code 'D'. The following day you return to start the next quadrat, and reach subquadrat 4,1 late in the day. There is a plant near the boundary that is not on your map, and since it must be a new plant, you place a tag on it. To double-check that you wrote down the new tag number correctly, you lift the new tag, and—surprise!—there is an old tag, 467, buried in the litter around the same plant.

This is a potentially insidious error, because if the mapper had not noticed the old tag, a single error (minor mismapping) would have become two major errors (a living plant called dead, and an already-tagged plant called a recruit and double-tagged). We have indeed found doubly-tagged trees at BCI. This sort of error cannot be detected by screening the database—there does not appear to be anything wrong! Thus, mappers have to check carefully whether each apparent recruit already has a tag and they have to search thoroughly for each dead plant. At the Luquillo plot, a metal detector was used during the recensus to help find fallen tags and thus improve the search for dead trees.

Errors in mapping provide an example of one of my overriding pieces of advice about the census work—think about the next census five years ahead! Accurate mapping during one census will reduce errors during the second. Near a boundary, only a few centimeters of error can land a plant in the wrong quadrat, so map especially carefully near boundaries.

Death

In the Mudumalai plot in India, trees are recorded as dead if the original aboveground tissue is killed, even if the root base forms new sprouts. These plants would be considered alive at BCI or Luquillo. There was a reason for the difference, though: fires are common at Mudumalai and many trees are killed above ground but maintain large roots which can sprout. This behavior is not known at BCI. I thus believe that mortality figures for the two plots are as comparable as they can be.

We decided against measuring dead trunks at BCI because the majority of dead trees either have no trunk or a rotting trunk from which a dbh would be irrelevant anyway. Moreover, since we do not know the date of death, any growth information from the dead trunk would be compromised. The only situation where I have regretted this rule is where allegedly dead trunks come back to life in a future census; in these cases, we have no dbh from the year when the tree was mistakenly called dead. Such errors are rare, however.

Locating the Prior POM

There have been a wealth of problems relating to re-measuring unusual stems in the BCI plot. Many of the rules that I recommended above and in chapter 2.2 have arisen because of these problems.

Big trees have been a big problem since the beginning. As noted in chapter 2.2, they were measured around the buttresses in 1982. Since then, measurements have always been taken above buttresses, but frequently not high enough above. This has caused confusion because multiple measurements have often been necessary—at the old POM plus at a new POM above the buttress. Painting the POM was not done on buttressed trees until 1990. Much confusion could have been avoided by following this simple rule: the POM should be 50 cm above the buttress and painted.

Multiple stems have been another source of problems at BCI, because through 1995, no individual stems were indicated. If one or more die, growth rates are problematic. Marking the large stem will obviate this problem in the future.

Resprouts have been an additional major headache at BCI, because the code 'R' has not been applied correctly. In 1990, stems which already had an 'R' in 1985 were invariably called resprouts again, because prior codes were not placed on the old data sheets. We thus had many 'R' codes in 1990 on plants which were not really resprouts. Rules for resprouts are probably the most difficult to follow, so they must be taught rigorously.

The resprout problem, as well as other changes in POM, convinced me that field sheets should have old codes and that mappers should make decisions in the field about changes of POM. It is easier for the field worker than it is for the database analyst months later. The two POM codes—'C' for a change in POM and 'A' for an alternate POM (not breast height)—are additional ways to eliminate the many problems that POM changes can cause.

Checking Dbhs

In 1995 at BCI, we extracted 1150 cases of extreme growth using the method described above—the most extreme positive and negative growth rates in several dbh classes. Of these, 241 seemed to be legitimate growth records (fast-growing trees), and 27 more appeared to be cases where the POM had changed, thus causing a large decline in dbh. The rest were apparent errors: 595 were obvious dbh errors made in 1995, and 255 must have been dbh errors made in 1990. Of the 595 errors from 1995, 323 were poor dbh measures and the rest were cases where the resprout code was forgotten.

The dbh re-measure exercise in 1995 (Fig. 2.4.1) showed that about 0.4% of the stems had errors of 50% or more in dbh. The 323 cases of egregious 1995 errors mentioned above represent about 0.15% of all stems in the plot. It thus appears that extracting the 1150 most extreme growth rates caught a substantial fraction of the extreme dbh errors. We thus hope to have reduced the frequency of large errors by an appreciable amount. These large errors haunt growth estimates.

I have no information from recensuses at other plots to compare with BCI. It will be especially interesting to find out whether plots in which all stems are painted—even those with the normal POM at breast height—show lower rates of error in growth.

Time and Labor

I have a precise estimate of the time it required to recensus the BCI 50 ha plot in 1995: a total of 13 people worked for nine months, then 3-6 people finished up odds and ends for three additional months. It took a total of 127 person-months. This broke down as follows: the main census proceeded at about 0.7 quadrats per person per day, requiring 85 person-months to complete. Checking quadrats and identifying recruits proceeded at about four times this rate, requiring 25 person-months for the field coordinators. Measuring the big trees required twelve person-months, and five person-months were used to correct dbh errors. This was a good crew, though—five of the field workers had worked on the 1990 census—so this may be a faster rate than other plots will achieve on their first recensus.

Key issues for the recensus

- Plants that look dead are not always dead
- Missing plants occasionally turn up a few meters from where they were mapped, sometimes in an adjacent quadrat
- Great care must be taken that apparent recruits do not have a tag, so that new tags are not placed on already-tagged plants
- Well-done maps in one census make finding trees in the next census far easier
- Multiple-stemmed plants are easy to handle if the main stem is indicated with the tag
- Buttressed trees are easy to handle if POMs are well above the buttresses and marked with paint
- Resprouted stems cause great confusion, and mappers should decide in the field whether a stem-break since the prior census invalidated the original POM
- Plants called resprouts in one census must not be called resprouts in the next census unless they break and sprout a second time
- Field sheets should have codes and dbhs from the prior census so that mappers can assess resprouts and other unusual situations
- The code 'C' allows mappers to indicate, in the field, all changes of POM.

Chapter 2.6

Affiliated Censuses at BCI

The large plot inventory provides thorough demographic information on trees, saplings, and shrubs larger than 10 mm dbh. Important phases of the life cycle—reproduction, fruit dispersal, and seedling growth and survival—are missed. Leaving these out of the main census was an entirely practical decision, given the high density of seeds and seedlings that can occur.

This chapter provides a brief description of long-term, ongoing censuses in the BCI 50 ha plot aimed at fruits, seeds, and seedlings. In addition, I describe canopy and liana censuses that complement some of the large plot projects.

Seeds and Fruits

In 1986, 200 seed-traps were placed near trails in the 50 ha plot (J. Wright, O. Calderon, personal communication). Each trap is a 0.7 x 0.7 m square of fine, flexible mesh, supported 1 m above the ground by four PVC tubes. Each week, all seeds and fruits in every trap are collected and identified to species. This has provided quantitative estimates of total fruit and seed production in the forest, seasonal changes in production, and dispersal information, since location of adult trees is known from the main plot census. One person completes the census in a year, but working just three days a week (the other two days are free for other work, such as data entry).

In forests where the flora is not as thoroughly-studied as Barro Colorado's, identification of seeds and fruits, often from fragments alone, presents a considerable difficulty. However, because seeds and fruits can easily be collected, sorted, and saved until a later date when identification is possible, a seed-trap project can begin immediately, even in the absence of prior collections. Because the traps provide a tremendous amount of information on tree fecundity and dispersal, they should become standard in every big plot.

Seedlings and Small Saplings

Two different seedling censuses are now being carried out in the BCI plot. One is associated with the seed-traps: around each, all saplings < 10 mm dbh but taller than 50 cm are censused annually in subquadrats of 5 x 5 m, and immediately adjacent to each trap, three 1 x 1 m squares are censused for all woody plants < 50 cm tall (J. Wright, pers. comm.). A hemispheric photograph is taken of the canopy above each trap, once each year, to estimate the quantity of light reaching the ground. The second seedling census is spread on a grid through a 5 ha region of the plot: all

Tropical Forest Census Plots: Methods and Results from Barro Colorado Island, Panama and a Comparison with Other Plots, by Richard Condit.

woody plants ≥ 20 cm tall but < 10 mm dbh are censused in a 1x1 m square immediately northeast of every 5x5 m stake in the region. Each of the two censuses includes lianas and trees, and has been carried out annually since 1994.

The census around seed-traps is completed in three months by one person plus an assistant, and the hemispheric photos require one person another month. The 5 ha census takes two people six weeks to complete (the latter is intended for all 50 ha, and would thus require 10 people about three months of work). Seedling censuses should be finished in the same three months every year to avoid seasonal fluctuations in seedling density. This is not so important for the census of plants ≥20 cm tall, most of which are over a year old, but smaller seedlings—those in their first year—undergo very large annual fluctuations which could confound a long-term census.

Seedlings are more problematic to census than seeds and fruits, since they cannot be collected for future reference (without killing them). At BCI, identification has been facilitated by prior work on seedlings, especially Garwood (1983). Similar work in other forests may require extensive background studies on seedlings.

Another difficulty with studies of seedlings in tropical forests is the very low rate of recruitment into the sapling size class. Mortality rates are high during the first year and mean growth rates exceedingly low (DeSteven 1994), so accurate estimates of the transition rate from seed to small sapling will require very large samples. Working out complete life tables that include this transition for common species, let alone rare species, may not be possible without much larger censuses.

Canopy Height

Every year since 1983, a profile of forest height has been taken every 5 m throughout the BCI 50 ha plot (Welden et al 1991). Immediately above each 5-m stake, a field-worker holds a telescoping, fiberglass measuring pole, extensible to 5 m. A leveling bubble is connected to the side of the pole to assure that it is exactly vertical. The imaginary line extending above the pole to the sky is divided into height segments above the ground: 0-2, 2-5, 5-10, 10-20, 20-30, and ≥ 30 m. A hand-held, focusing range-finder is used to estimate height above the pole. Vegetation is recorded as either present (if it contacts the line) or absent, in each of the six segments. There are 20,301 points censused in the 50 ha plot (every 5-m stake within the plot and on all the boundaries).

One difficulty that has arisen is how to account for deciduous species. The field-worker is taught which species can be deciduous and is supposed to always record canopy height as if these species had leaves. That is, if the vertical line passes through the crown of a leafless tree that is judged to be alive, it should be counted as "vegetation present", but not if the tree is dead.

The canopy height data have been used to map canopy gaps and their changes through time (Hubbell and Foster 1986b) and as a way to assess light environment for saplings below (Welden et al 1991). The canopy census was done once at Pasoh, but never at any other large plot. In very tall forest, such as those in southeast Asia, a taller height category should be added, 30-45 m, then 45 m and above.

Completing the canopy height census requires one person eight months at BCI.

Lianas

Lianas have been included in the large plot census at two sites, Sinharaja in Sri Lanka and Ituri in the Congo. At Sinharaja, diameter was taken 1.3 m from the rooting site, measured along the stem, and lianas were included only if the diameter exceeded 50 mm. At Ituri, the point-of-measure was where the stem first reached 1.3 m vertical height above the ground, and plants were included when the dbh exceeded 20 mm. Although the two sites will yield basic data on liana density and diversity, the lack of standardization precludes comparability. I suggest that other plots considering liana counts begin with the Sinharaja method, including smaller diameters if possible.

Altogether, the seed traps, the seed and seedling censuses, liana censuses, and the main 50 ha plot data set provide precise demographic information on every life history phase of woody plants, from seeds to the largest adults. Complete life tables can be created for large numbers of species, and these can be coupled with the estimates of light environment provided by the canopy profile and the canopy photos. No other tropical forest research program has come near such complete demographic coverage of a community.

CHAPTER 2.7

Summary and Overview

I have two different ways to conclude this section on field methodology. First, I provide a summary table which compares and contrasts the methods used at the 12 large census plots. This highlights the important techniques used and rules followed, where they are the same and where they differ between plots. Second, I collect all my estimates of the time, labor, and equipment needed and come up with an estimate of what it takes to complete a large plot.

Similarities and Differences

A quick overview of the 12 large plot sites confirms that the methods used have been standardized reasonably successfully (Table 2.7.1). All the plots map all stems ≥ 10 mm dbh using a carefully-established grid of 20 m and a finer grid of 5 m. Rules for measuring dbh are remarkably consistent at all sites, thanks especially to the methods pamphlet published by Manokaran et al (1990) and another written in Spanish for the Ecuadorian and Panamanian plots (never published). The only major variant in dbh measurements has been at Pasoh and BCI, where figures were rounded down to 5 mm intervals in early censuses (see chapter 2.2); both plots have abandoned this practice and now use millimeter accuracy, as at other plots.

Plot size and shape varies somewhat, with the Ituri forest being most divergent. No plot is smaller than 16 ha (Table 2.7.1). Other variation in methodology between sites govern palms, stranglers, and mortality of resprouting individuals (Table 2.7.1). The Luquillo plot had a more liberal rule for including palms, and palm density and overall density, and thus is not strictly comparable with other plots (since one palm, *Prestoa montana*, is abundant there), however, the difference is not great (chapter 2.2). Stranglers were included at some sites and not others (Table 2.7.1), but stranglers are rare.

Another discrepancy was the way mortality was defined at Mudumalai, where stems which died above ground but resprouted from roots were counted as dead. Resprouts at BCI were not counted as dead. This lack of precise standardization in this instance originates from a fundamental biological difference—fire at Mudumalai—and seems unavoidable.

A final difference in technique has to do with multiple-stemmed plants. Every plot actually follows the same rule of counting clearly connected stems as a single individual, however, separate stems receive separate tags at some plots but not others (Table 2.7.1). This should cause greater errors in growth estimates at BCI, compared to Mudumalai or Ituri, but there should be no bias. Growth rates should be comparable among the sites.

Tropical Forest Census Plots: Methods and Results from Barro Colorado Island, Panama and a Comparison with Other Plots, by Richard Condit.

Table 2.7.1. Key methodological similarities and differences across the CTFS network of large plots. Informtion from the plot in Columbia was not yet available when this table was assembled. Resprouts are only an issue at the two plots where recensuses have been finished.

site	grid	dbh limit	lianas included?	multiple stems	resprouts alive?	palms included?	stranglers included?
BCI	20 m & 5 m, permanent	≥ 10 mm	no	not separated	always alive	with stem at breast height	when host gone
Luquillo	20 m & 5 m, permanent	≥ 10 mm	no	not separated		when height > breast height	when host gone
Yasuní	20 m & 5 m, permanent	≥ 10 mm	no	largest painted		with stem at breast height	with trunk ≥ 200 mm
Pasoh	20 m & 5 m, permanent	≥ 10 mm	no	not separated		with stem at breast height	never
Huai Kha Khang	20 m & 5 m, permanent	≥ 10 mm	no	not separated		with stem at breast height	never
Lambir	20 m & 5 m, permanent	≥ 10 mm	no	each tagged		no	when host gone
Palanan	20 m & 5 m, permanent	≥ 10 mm	no	not separated		with stem at breast height	never
Mudumalai	20 m & 5 m, permanent	≥ 10 mm	no	each tagged	some dead [1]	none present	when host gone
Sinharaja	20 m & 5 m, permanent	≥ 10 mm [2]	yes	not separated		with stem at breast height	never
Korup	20 m permanent, 5 m during census	≥ 10 mm	no	largest tagged		none present	none present
Ituri	20 m permanent, 5 m during census	≥ 10 mm [2]	yes	some tagged		none present [3]	none present

[1] Considered dead if all above-ground stems died, but alive if at least some above-ground tissue survives

[2] Lianas ≥ 50 mm dbh included at Sinharaja, ≥ 20 mm dbh at Ituri; free-standing stems ≥ 10 mm included at both plots

[3] There are two palm species in the census, but they are lianas and would not be included at other plots

Labor Needs

Most of the expense in completing a large plot is labor. Table 2.7.2 assembles estimates of the labor needed to finish the topographic map, the main mapping and identification phase, and a recensus. High and low estimates are given for the mapping and identification phase, since different plots have required rather different amounts. These estimates apply to plots with about 5,000-7,000 trees per ha, which is the density found at most of the large plots currently underway. The Mudumalai plot, however, has only 400-500 trees per ha and has required far less work, and the Huai Kha Khaeng plot has only 2,000 trees per ha.

Table 2.7.2 does not include labor by chief scientists who develop and oversee the project. All of the large plots have at least one scientist who spends a considerable amount of time coordinating activities, and some have three or more, but it is impossible for me to factor in their time (except where they are heavily involved in field work, and that is included in the table). Salaries for principal-investigators are seldom considered part of a project's cost anyway.

Equipment Needs

The major equipment and supplies needed to finish a large plot include the surveying compass, stakes for marking the grid, tags, a tag puncher, tag string, flagging, paper, and computers. Costs for these items are estimated in Table 2.7.3, giving a total of $56,000 for a 50 ha plot. Other substantial items may be necessary, though. At most large plots, for example, a truck is necessary for transporting the field workers, and boats may be necessary in some places. There are a variety of other minor supplies which I have not attempted to include, such as pencils and computer disks, and other equipment and supplies will prove necessary at some plots, but not others—camping gear, snake-bite kits, etc. The total bill for equipment might thus be $20,000-40,000 more (mostly for cars or boats). On the other hand, there are substantial cost saving measures that have proved acceptable at some of the plots. Over $10,000 was saved by using stones instead of stakes for markers at Mudumalai, and $15,000 was saved by using home-made tags in Cameroon (both savings rely on cheap local labor, however).

Time

At BCI and in Ecuador, 12-14 people work on the censuses at one time, but in Sri Lanka and the Congo, 15-25 people work in large teams. Obviously, the number of people that can be put to work determines how long it will take to complete a large plot, but most of the plots completed to date have taken around three years. A recensus can easily be done in a year.

Total Cost for a Large Plot

Labor will always be the bulk of the cost for a large plot, and since labor costs vary tremendously from country to country, the total cost of a plot can vary widely. It also depends on whether there will be station fees or food costs for the field workers which must be born by the project, as will usually be the case if workers must stay at a remote field camp.

The 50 ha plot in Yasuní, Ecuador, will end up costing about $500,000: although labor is moderately inexpensive, we must pay $20 a day station fees and transport costs for the field workers, and the labor in identifying all the specimens is much greater than at other plots. We recently prepared a budget for a 50 ha plot in

Table 2.7.2. Estimates of labor needed to complete two censuses at a large, tropical forest plot. Data entry is covered in chapter 3.1, but the labor for this is included here.

Phase of work	Person-months/50 ha	Person-months/ha
Topography	30	0.6
Mapping and identification	375-500	7.5-10
Data entry (double)	28	0.6
Recensus	127	2.5
Recensus data entry	14	0.3

Table 2.7.3. Costs of equipment and supplies needed to complete two censuses at a large tropical forest plot. See chapters 2.2 and 2.3 for more information on suppliers for some items. Asterisks indicate items for which use of home-made goods can lead to considerable savings, as long as local labor is inexpensive (see text). Shipping costs are obviously highly variable. Note that the table does not include vehicle costs, and many plots require full-time vehicle use.

Item	Cost/unit ($US)	Units needed	Total cost ($US)
Surveying compass	1,500	1	1,500
PVC stakes for 20 x 20 grid*	1.20	1,400	1,680
PVC stakes for 5 x 5 grid*	0.33	21,000	7,000
Tags* (unit=tree)	0.05	400,000	20,000
Tag string (unit=tree)	0.007	400,000	2,800
Tag punching machine	1,375	1	1,375
Computers, printers	2-4,000	4	12,000
Waterproof paper (unit=sheet)	0.085	24,000	2,040
Plain paper (unit=sheet)	0.009	24,000	216
Calipers	25	10	250
Dbh tape	30	10	300
Colored flagging (unit=meter)	0.022	20,000	444
Compass	8	10	80
Binoculars	100-300	2	400
Clipboards, nails, hammers, rope, tape measures, ladder	—	—	~1,000
Shipping	—	—	~5,000
Total			$56,185

Amazonas, Brazil, and this was also about $500,000. On the other hand, for a 25 ha plot in La Planada, Colombia, less than $125,000 was budgeted because local labor is inexpensive and people can commute from their homes to the site. As a rough starting point, I suggest that the least expensive 50 ha plot will cost about $250,000, where labor and station fees are cheap. Many sites will cost about $400,000, and rather more expensive and difficult sites $500,000.

Part 3: Database Methods for a Large Forest Plot

Part 3

Introduction and Overview

In this section, I provide detailed methods for constructing and managing the data generated by a large forest plot. This is an important and perhaps overlooked stage. A well-designed database will make data management easy and efficient: it facilitates error-correction as well as subsequent demographic analyses. On the other hand, mistakes during database construction can lead to new errors; in fact, many of the most frustrating and difficult-to-solve problems at Barro Colorado resulted from database errors, not field errors. I offer here advice on avoiding these errors.

The first step in data management is simply to get the information recorded on field sheets entered into a computer database, and this is covered in chapter 3.1. Chapters 3.2 and 3.3 then explain methods for organizing this preliminary database into a finished format convenient for data analysis. Re-formatting the database in this way is particularly useful after a second census, and is designed to make all the crucial information about every individual clear, concise, and easily accessible in the computer. These chapters are accompanied by computer code we have developed at Barro Colorado for database construction. Most are written in FoxPro's programming language, since this allowed the data to be entered directly from field sheets into a FoxPro database. Readers wishing to take full advantage of the programs will have to learn some FoxPro. One of the programs is in Systat's graphics module, Sygraph.

One last chapter in the database section covers a completely different topic. Large projects involve many scientists, and it is important to spell out intellectual property rights to the data. Chapter 3.4 describes agreements we have used for the Barro Colorado plot.

In the database section, I change the presentation relative to the field methodology section in that I do not offer alternate approaches after describing a recommended approach. Obviously, there are alternatives, but I felt uncomfortable presenting them for two different reasons. First, describing the database management requires intricate details on data structure and computer programming, and repeating alternate methods would have expanded and complicated the presentation a great deal. Second, data management at Barro Colorado has evolved to a much more sophisticated level than it has at other large forest plots—indeed, only 5 of the 12 large plots have finished databases as of 1998. Thus, there are not many alternatives in use at this moment.

Tropical Forest Census Plots: Methods and Results from Barro Colorado Island, Panama and a Comparison with Other Plots, by Richard Condit.

CHAPTER 3.1

Data Entry

In the not-too-distant future, field workers in the forest will use hand-held computers to record the information they collect, and after a few days of work, results will be downloaded directly into a large computer. For the time being, though, field data are recorded by pencil and must be transcribed later into a computer for analysis. However, data entry is more than just copying field sheets into a computer: there are bigger issues to consider. First, it is crucial to catch and eliminate errors during data entry and database organization. This is especially true of propagating errors—those that multiply by causing further errors when databases are linked. Second, data entry must anticipate database organization: the separate databases created from various field sheets must be easily merged into a single, well-organized unit that fosters analysis.

This chapter covers the first phase of getting the data into the computer—creating a series of preliminary files containing all information collected in the field. Computer programs for data entry and error correction, written in FoxPro, are provided. A large number of files are created along the way, and these are merged, matched, and checked at several stages. To help keep track of the changes, Figure 3.1.1 shows a flow-chart of database work, summarizing the steps that are described in detail below. Chapters 3.2 and 3.3 continue with database assembly, explaining how the preliminary files created here are merged into an analysis-ready final database.

Initial Data Entry

Data Files Created

For a first-time census, data for each 20 x 20 m quadrat are entered into the computer in three separate files, one for each of the three field sheets used per quadrat: the main data sheet (Fig. 2.1.B), the multiple-stem data sheet (Fig. 2.1.C), and the map (Fig. 2.1.A). Big trees, those measured by special teams after the main census, are entered in a fourth file. For a recensus, there is one more file, since the main data appear on two data sheets—one for recruits and one for original trees (see chapter 2.5).

The Main Data File

Most information about each tree comes from the main data sheet, and is entered into what I call the main data file or main database. The most important data are quadrat number, subquadrat number, tag, dbh, species name, and codes. All are

Tropical Forest Census Plots: Methods and Results from Barro Colorado Island, Panama and a Comparison with Other Plots, by Richard Condit.

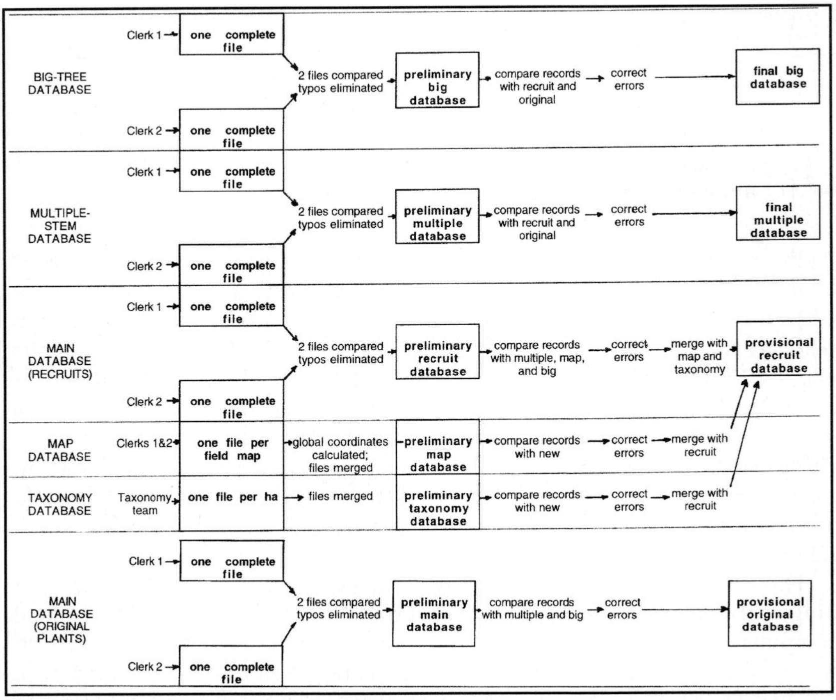

Fig. 3.1.1. Flow-chart for data entry and preliminary database assembly. FoxPro data files are in boldface. The data files in the far right column are referred to as provisional because they form the basis for further database manipulation, described in chapters 3.2 and 3.3. For a first-time census, five major files are created. For a recensus, an additional file is necessary for original trees, those already tagged in the first census and re-measured in the second. Recruits in the second census are treated just like all trees in the first census.

entered exactly as they appear on the data sheets. The species name will be blank if the taxonomy team worked separately and created their own database. It is then left blank by the data-entry clerks.

There is additional administrative data for each quadrat that goes in the main database. These appear at the top of the data sheets: the dates of the original census, of field checks (if different), and of computer entry, and the initials of field and clerical workers who worked on the quadrat. This information will be entered and stored for every record in the initial database (but, as explained more fully below, data-entry programs provide short-cuts so that duplicate data need not be re-typed). Most of the dates and names have strictly administrative uses—to know who did what, when, and to chart progress of all phases of work. Most will eventually be moved into a separate file and deleted from the final database (chapter 3.2). The exception to this is the date of the original census, which is needed in many analyses; it will remain in the final database (chapter 3.2).

Recensus

Data on recruits during a recensus—newly encountered plants tagged and mapped for the first time—are entered into a recruit database. This is identical to the main database from a first-time census, described above. Indeed, all trees in a first time census are recruits.

But most measurements made during a recensus are on trees that were already tagged, and these were entered on a separate data sheet for original trees (chapter 2.5). These data will be entered into a separate database, distinct from the recruit database (Fig. 3.1.1). In most ways, this original-tree database is the same as the main database from a first-time census or the recruit database, however, less information need be entered for every tree. Species name, quadrat, and subquadrat are unchanged from the first census, and thus already in the computer. Only new data—dbh and codes—must be added. Dates of census and names of personnel also go in the original-tree database.

Multiple Stems and Big-Tree Files

This information was recorded on separate data sheets, whether during a first census or a recensus, and is entered into separate databases. In the multiple-stem database, each dbh is entered into a separate record, so that a tree with three stems appears on three different lines. The tag number is the same on all three, of course. The database for big trees is equivalent. In a first census, a single dbh would be entered for each plant, but after a second or third census, some big trees may have additional dbhs because buttresses ascend and force higher points-of-measure (see chapter 2.5). These are handled just like the dbhs of multiple-stemmed trees.

Double-Entry

All of these databases should be independently entered by two different people; this is referred to as double-entry. This eliminates typographic errors made during the data-entry process: the duplicate files are compared entry-for-entry, and discrepancies indicate errors in transcribing from the field sheets.

Specialized Data-Entry Programs

These databases are created and filled using programs specially designed for use with each of the data sheets (Programs 3.1.1, 3.1.2, and 3.1.3). Program 3.1.1 is for creating the main database for a first census, or a recruit database from a recensus (because these are identical). Program 3.1.2 creates the original-tree database, and is thus only used with a recensus, while Program 3.1.3 is for multiple stems and big trees from any census.

The programs display a FoxPro database that matches each field sheet. As data are entered, the programs move to subsequent blank entries. Most useful, the programs automatically fill in repetitive information, minimizing the amount that has to be entered by clerks. For the first record from a quadrat, the clerk enters the dates, names of personnel, quadrat number, then the information specific to the first tree (subquadrat, tag, dbh, codes). When he or she presses the plus key on the numeric pad, all dates, names, and the quadrat number immediately appear on the second line, and the cursor jumps to the subquadrat field. From then on, just the four fields need to entered. When a new quadrat is started, each of the preliminary fields must be edited, but then the new values are copied downward as that quadrat is finished (Box 3.1.1 provides more details).

The programs also include some simple data-screening steps to help catch errors. If illegal entries are attempted, the programs pause and display a message on the screen (Box 3.1 gives details). Some illegal entries are simple typos that the clerk can immediately correct. But illegal entries will also occur when the data sheet has an error. If so, the clerk should place a –1 in the field in question. On a regular basis (perhaps once a week, depending on how many there are), the database manager can extract records with –1 and try to resolve them all. If the data sheet was in error though, the –1 remains in the database, and a list of all such problems must be sent to the field coordinators to be resolved (perhaps once a month depending on the error rate).

When a clerk starts work every morning, an existing database is opened and new records appended at the bottom. A single large database, expanding as entry proceeds, is created. It is better to have clerks work for about 3 weeks, turn over their file to the database manager for checking, then start a new file. This keeps the database sizes smaller and hastens data entry. At each 3-week break, the database manager should compare the double-entered duplicate files and eliminate typographical errors. A FoxPro program for comparing duplicate files is given as Program 3.1.4. It copies all non-matching records to a new data file, and the entry clerks then consult the data sheets and make the appropriate corrections in their databases (Box 3.1.4). Double-entry correction generally requires several rounds of comparison (i.e., comparing, fixing errors, comparing again, etc.).

After correcting each 3-week file, the database manager merges it with earlier files, thus maintaining an expanding database for the entire plot. When completed, there is one large main database, one multiple-stem database, and one big-tree database. For a recensus, instead of the main database, there are recruit and original databases.

Details on how these programs work are provided in Boxes 3.1.1-3.1.4; the source code is given in full in Programs 3.1.1-3.1.4. The programs are written in FoxPro version 2.5, and a thorough understanding of FoxPro databases, commands, and programming is needed to understand the source code and how to manipulate it. However, the programs can be copied and run without understanding the source code, and minor changes are easy. Using FoxPro for data entry allows the data to be assembled directly into a FoxPro database, and this allows a variety of data screening analyses to be run immediately, even before the data are fully entered (see chapter 2.4).

Field Maps

Maps are entered differently from the other field sheets. First, double-entry is not necessary, since important errors made while digitizing will be detected by other checking routines (as described below). Second, map data are entered into one data file per field sheet, rather than one large, expanding file which includes all data.

At BCI, we digitize maps using a 30x43 cm digitizing tablet by Jandel (model 2210-1217, resolution 0.001 inches). Three corners of each 10x10 m map are spotted with the reticle, and then the location of every new tree as well; the tag number is entered after clicking on the tree location. Detailed instructions come with the digitizing tablet. A simple macro written with Sigma Scan software creates an ASCII data file holding tag number and local x-y coordinates within the quadrat. These coordinates give the perpendicular distance of each tree from the lower left corner of the map, and are thus always < 10 m. I refer to these local coordinates as *lx* and *ly*.

Data from each 10x10 m map are saved to a single file. Since there are four maps per 20x20 m quadrat, there are four map files per quadrat. Each file is named with the quadrat number and a second number, one to four, to indicate the 10x10 m subregion. The first map from quadrat 3513, for example, includes the lower left 10x10 m block, subquadrats 11, 12, 21, and 22 (chapter 2.2), and the associated map file is called Q3513.1. The next map file from the same quadrat, with subquadrats 13, 14, 23, and 24, is named Q3513.2, etc. (Fig. 3.1.2). After finishing a 50 ha plot, with 1250 quadrats, there are 5000 map files.

The next step is merging the 5000 map files into a single FoxPro database. The name of each file, such as 3513.1, must be incorporated as a new field in the large database (otherwise the quadrat number is lost). The "global coordinates"—the location within the entire 50 ha plot—are calculated from the local coordinates within each 10x10 m region and the quadrat number, and are stored in the same FoxPro database. I refer to global x and y coordinates as *gx* and *gy*, and they are the basis of all spatial analyses. A FoxPro program for converting the individual quadrat files into a large map database is included here (Program 3.1.5). Box 3.1.5 documents the program and explains the calculation of global coordinates.

Merging should begin before all 5000 maps are completed, though. I suggest that maps be merged every 3-4 months, when error-screening takes place (see next section). Thus, after the first 10 ha, all maps are merged into a single file which is checked against the main file. Any errors found are corrected in the original map files, which are maintained as digital files. After a subsequent 10 ha are digitized, the maps from the first 10 ha are merged with maps from the next 10 ha, and data-screening repeated.

During a recensus, when 20x20 m maps are used in the field, just one file is created per quadrat. It can be named simply Q3513, for instance, without the extension .1 or .2, etc. Only the recruits are mapped, since the original trees were mapped in the prior census. Thus, there are fewer files, just 1250 in 50 ha, and each file is much smaller. Still, the procedure for calculating global coordinates and merging the small files into one large one is essentially the same as for the first census (Box 3.1.5).

The Taxonomy Database

Where the flora is complex and not well-known, the taxonomy teams should work separately from the mapping teams, and they will not use the main data sheets (chapter 2.3). The taxonomists construct their own separate database by copying the main database (MAINDATA.DBF) entered with Program 3.1.1. From it, all fields except the tag, quadrat, subquadrat, mnemonic (or species code), and dbh should be eliminated. New fields for full species name, field notes, prior species name, and an indication of whether the plant has been collected should be added (see chapter 2.3).

Taxonomy work begins as soon after the mapping as possible, so the main database will not be complete. Whatever portion is available should be copied, for example, after the first two hectares are entered in the database. When these are completed, additional portions of MAINDATA are copied as needed, then appended to the taxonomy data file.

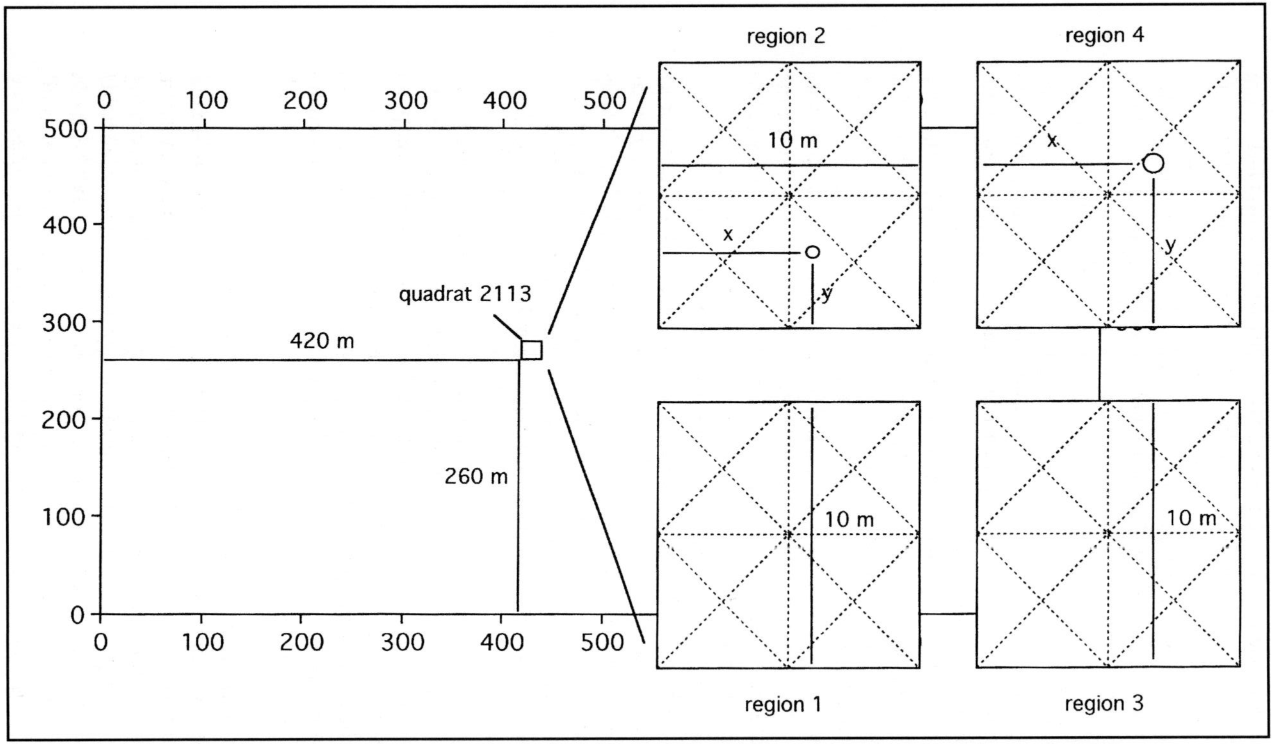

Fig. 3.1.2. A map of a 50 ha plot illustrating the conversion from local coordinates on a field map to global coordinates within the plot. The right-hand portion is a blown-up diagram of a single quadrat, which is mapped in the field on four data sheets. These are numbered regions 1, 2, 3, and 4, as described in the text. The tree mapped in region 2 at coordinates x, y, has coordinates $x, y + 10$ for the entire quadrat. The left-hand portion of the figure locates the quadrat within the whole plot. As described in chapter 2.2, quadrat number 2113 means column 21, row 13, where columns and rows refer to 20-m sections. Since the leftmost column is column 0, column 21 starts 420 m from the left edge, and likewise row 13 starts 260 m from the bottom edge of the plot. The global x-coordinate (*gx*) of a plant in quadrat 2113 is 420 plus the x-coordinate within the quadrat, and the global y (*gy*) is 260 plus the local y.

In some plots, for example where the flora is well-known, taxonomists work with the mapping teams. Species names are filled in on the main data sheet, and there is no separate taxonomic database. The names are entered by the clerks along with the rest of the main data, using Program 3.1.1. The same approach is used for recruits during a recensus; species names are entered with the rest of the main data.

Checking the Data

As the main, multiple-system, big-tree, map, and taxonomic databases are enlarged, screening for certain obvious errors should begin. I suggest running the following routine data screens every 3-4 months. The same advice applies in a recensus for the original and recruit databases. Table 3.1.1 lists some mistakes which can easily be uncovered by simple screening.

Non-Corresponding Records

The main, map, and taxonomic databases must have exactly the same list of plants—one record for each individual should appear in each file. Non-matching records *must be* corrected before merging. For example, it is likely that the map file will be missing records found in the main file: cases where trees were missed while digitizing the maps. In addition, every tree with a code "M" in the main database must appear in the multiple stem database, and vice versa, and likewise for trees with code "B" and the big tree database. Every 3-4 months, databases should be compared against each other, and all non-matching records identified.

This is easily done in FoxPro. Using the *SET RELATION TO* command, any pair of databases can be linked through their tag numbers. In the linked file, any record missing in one file shows up as a blank tag. (To compare the multiple-stem and main databases, it is first necessary to extract from the latter all records with code "M", then link those with the former.) FoxPro can copy all records with blank tags to a new database, one for each pair of matched files. The complete list of non-corresponding records is printed, and data clerks check each against the data sheets and fill in missing records as needed. In some cases, the missing data will not be on the data sheets, indicating that the field workers forgot to record it. The clerks should save these errors in a master list. These will be combined with other errors requiring field work, eventually to be given to the field coordinators to be resolved.

Duplicate Records or Tag Numbers

Duplicate entries in the database are evident when a tag number appears in either the main, map, or taxonomy database more than once. This is a critical error, and the duplicates must be removed before merging the different files. If they are not, then a record in the main file might be linked with the wrong record in one of the other files. In some duplicates, an entire record appears twice, with all fields identical, simply due to a typing error. The more difficult type of duplication is when the same tag number appears twice but with different data. This is usually a field error, where the same tag number was recorded for two different plants.

Program 3.1.6 searches for duplicates by indexing a database on tag, then checking whether consecutive indices have the same tag (Box 3.1.6). The output is a database of all duplicate tags, each of which must be checked by the clerks. Some are resolved immediately, but others require attention in the field and must be appended to the list of errors for the field coordinators.

Missing Entries

All records with –1 in any field should be extracted and printed to a separate file. These represent problems found during data entry. In addition, any blank records that should not be blank should be included in the list (exceptions are codes and the dbh of dead trees, which can be blank). Blanks should have been noticed during data entry, with a –1 entered, but some may have been missed. As with other errors, FoxPro is used to copy all records that are blank or –1 to a new database. After checking against the data sheets, unresolved errors are appended to the error database for the field coordinators.

Table 3.1.1 summarizes common database errors encountered during data entry at BCI.

Back to the Field

The various errors collected by screening are merged into a master error database, and this is sorted by quadrat and printed for use as data sheets by the field coordinators. Corrections can be entered directly on the printout, but blank maps will also be necessary for plants that were omitted from the original maps. If a new dbh is recorded for a plant during the correction, the new date must also be recorded, and the date of census in the main database must be changed; for other corrections, such as location or codes, the original census date remains.

A record of all errors corrected in the field must be maintained, for each signals a discrepancy between the field sheets and the final database. This is important if, for example, future problems arise with a particular plant: the original data sheets plus the corrections would have to be consulted in order to understand the full history of the difficulty.

Randomly Re-Measuring Trees

In chapter 2.4, I suggested that each field team re-measure trees in one quadrat that a different field team had already completed. When the re-measurements are completed, they should all be entered by the clerks into a new database. The same entry programs (Programs 3.1-3.3) can be used, but the file name is different. When completed, this file is related to the main file so that the two dbhs can be compared. These data illustrate the degree of accuracy of dbh measurement, but they also indicate some records that need correcting. For all discrepancies ≥ 5%, the plants should be measured again—a third time—and the dbh in the main database corrected as needed.

Frequency of Error Correction

These various procedures for screening errors should not wait until the entire census is finished. Full database checks should be done roughly every three or four months: this means matching the doubly-entered files, merging all the map files, running the various error checks described above, and sending a list of errors to the field coordinators.

Merging the Databases

When the databases are completed and the final error corrections are made, the main data file, map data file, and taxonomy data file have records for exactly the same plants, and they should be merged into one large file (Fig. 3.1.1). The species designation is taken from the taxonomy database and put in the mnemonic

***Table 3.1.1.* Errors commonly seen in a large plot dataset that can be readily found by screening the data. Here, the main database refers to either the recruit or original tree database in a recensus.**

Error	Common cause	Revision
Typographic error	Clerk error	Resolved when comparing double-entered files
Location in map file does not match subquadrat in main database	Field worker either mis-mapped or mis-entered subquadrat number	Often resolved without field check
Plant in multiple-stem (or big-tree) database, but no 'M' (or 'B') in main database	Field worker forgot 'M' (or 'B')	Often resolved without field check by adding missing code
Plant with 'M' (or 'B') in main database but not in multiple-stem (or big-tree) database	Field worker forgot to fill in data	Requires field check
Duplicate tag (on different plants)	Field worker forgot to put tag on one plant but enters number on datasheet; places the tag on next plant	Requires field check
Plant in main but not map database	Field worker forgot to map tree, or clerk missed the plant on map	Can be resolved in lab if missed by clerk, otherwise requires field check
Plant in map but not main database	Field worker forgot to enter tree on main sheet	Requires field check
Illegal entry in any field	Writing errors in field or by entry clerk	Can be resolved in lab if missed by clerk, otherwise requires field check

field from the main database, but this is unnecessary if there is no taxonomy database. The gx and gy coordinates are added as new fields to the main database. These steps are carried out in FoxPro by linking the various databases by tag number. All the data fields are copied into a single large file. The multiple-stem and big-tree are kept separate for now, and will be discussed further in chapter 3.2 (Fig. 3.1.1).

In a recensus, there is also a separate file of original trees which do not appear in the map file, and this is also kept separate for now; merging it into a multiple-census database is described in chapter 3.3. During recensuses, there is no taxonomy file, as the species were identified during the main mapping phase.

Equipment, Time, and Labor

Computers

Data entry does not require sophisticated or high-powered computers, but efficient double-entry requires that two machines be available, and each data set requires about 30 megabytes of storage. Older PCs, such as 286 or 386 models with 100-megabyte hard drives, can suffice, although faster machines are desirable.

Once the data are entered, any routines that require reading all records—simple file merges and data-checking—will be tediously slow on slower, older computers. In addition, various steps require duplicating 30-megabyte files, so substantial hard disk space is useful. Thus, at least one PC with a Pentium processor should be available for merging and checking, to go with two less powerful data-entry computers.

Labor

One person can enter 50 ha of data—assuming approximately 250,000 plants—in 15 months: eight months to digitize all the maps, and seven months to enter the other data sheets. Double-entry of the main files and single entry of the maps thus takes two people less than a year. This is faster than the field work, which takes three years or more for 50 ha. Thus, data entry can be delayed some, but since taxonomy work and field checks profit greatly by having computerized databases, it should not be delayed long. Inevitably, data entry clerks will stay ahead of the field workers and thus often have no data to enter; it is best to have other work for them (copying and checking field sheets, or other administrative assistance with the plot).

Double-entering the recensus data at BCI has taken two people about seven months. Since field work on a recensus takes only about a year, it is possible to arrange a much more efficient interchange between data collection and entry. Data entry can start five months into a recensus, so that field work and data entry finish at about the same time. This still allows field checks based on the computerized database to be done in a timely fashion.

Data management of the growing computer databases also requires oversight by a data manager, however, this is not require on a full-time basis. Running the verifications, assembling and printing out errors for field workers, building the taxonomy database, and other odds and ends required about 20% of the BCI data manager's time.

Key Issues for Data Entry

- Double-entry eliminates typographical errors
- Data-entry programs speed entry by matching field sheets, checking for illegal entries, and automatically filling in repeating fields
- Clerks enter –1 for problems they cannot solve
- The database must be screened for obvious and important errors: duplicate tags, non-matching records in different files, missing data
- A database of errors is created for field coordinators to use in the field for making corrections; this and the corrections must be saved for future reference
- Main, map, and taxonomy databases are merged into a single large database.

Programs

Program 3.1.1A. Data entry program for the main datasheet, and the recruit datasheet for a recensus.

Program MAIN.PRG

```
CLEAR
CLEAR ALL
SET CONFIRM ON
SET BELL OFF
SET TALK OFF

SET PROCEDURE TO PROC
PUBLIC key_string, field_string

mN='  '
mF=ctod('  /  /  ')

DO prelim

SELECT 0
USE SPPLIST
SET ORDER TO MNEM

SELECT 0
USE MAINDATA
GOTO BOTTOM

start_fld=10
field_string=FIELD(1)
N=1
DO WHILE N < start_fld
    field_string=field_string+','+FIELD(N)
    N=N+1
ENDDO
SET CARRY TO &field_string

key_string=REPLICATE('+chr(9)', 5)
PUSH KEY
ON KEY LABEL + DO Plus

IF RECCOUNT()=0
    APPEND BLANK
ENDIF

```

Program 3.1.1A. (continued)

```
REPLACE DATAENT1 with mN, DATE_ENT1 with mF

BROWSE LAST NOWAIT FIELDS ;
  Q20     :V=VAL(SUBSTR(q20,1,2))>=0 AND VAL(SUBSTR(q20,3,2))>=0 AND ;
               VAL(SUBSTR(q20,1,2))<=49 AND VAL(SUBSTR(q20,3,2))<=24, ;
  NAME    :P='!!!' ;
          :H='INIT', ;
  Date    :B=CTOD('01/01/95'),CTOD('12/31/95') ;
          :H='DATE', ;
  Chq     :P='!!!' , ;
  Date_chq   :B=CTOD('01/01/95'),CTOD('12/31/95') ;
          :H='Dt_cheq', ;
  P5      :V=p5>='11' AND SUBSTR(p5,1,1)<='4' AND SUBSTR(p5,2,1)<='4', ;
  Tag     :P='999999', ;
  Mnem    :P='!!!!!!' ;
          :V=spLookUP(), ;
  Dbh     :B=-1,4000 ;
          :H='Diam', ;
  Code    :P='!!!!!' ;
          :H='CODES' ;
          :V=SUBSTR(code,1,1)$[ BDMRNTCSLQV.] AND ;
              SUBSTR(code,2,1)$[ BDMRNTCSLQV.] AND ;
              SUBSTR(code,3,1)$[ BDMRNTCSLQV.] AND ;
              SUBSTR(code,4,1)$[ BDMRNTCSLQV.] AND ;
              SUBSTR(code,5,1)$[ BDMRNTCSLQV.]
```

Program 3.1.1B. Procedures used by the data entry program for new trees. These are also used by the programs for entering data on old trees (Program 3.1.2) and multiple stems (Program 3.1.3) and need not be changed for those.

Program PROC.PRG

```
PROCEDURE prelim

  @6,18 SAY "Please enter the following information"
  @8,15 TO 16,60
  @11,27 SAY "Name : " GET mN PICTURE '@!A'
  READ
  @13,27 SAY "Date : " GET mF RANGE ctod('01/01/95'),ctod('12/31/95')
  READ
  CLEAR
  RETURN


PROCEDURE spLookUp

```

Program 3.1.1B. (continued)

```
    mFieldName = VARREAD()
    mMnem = UPPER(LTRIM(&mFieldName))

    mDataabbr = LOOKUP(spplist.mnem, mMnem, spplist.mnem)

    IF EMPTY(mDataabbr)
        ? chr(7)
        mDataabbr=REPLICATE('?', 6)
        DO speclist
    ENDIF

    REPLACE &mFieldName WITH mDataabbr

    RETURN .T.


PROCEDURE speclist

    #REGION 0
    REGIONAL m.currarea, m.talkstat, m.compstat

    IF SET("TALK") = "ON"
        SET TALK OFF
        m.talkstat = "ON"
    ELSE
        m.talkstat = "OFF"
    ENDIF
    m.compstat = SET("COMPATIBLE")
    SET COMPATIBLE FOXPLUS

    m.currarea = SELECT()

    SELECT SPPLIST
    GO TOP

    IF NOT WEXIST("specscrn")
        DEFINE WINDOW specscrn ;
            FROM INT((SROW()-19)/2),INT((SCOL()-18)/2) ;
            TO INT((SROW()-19)/2)+18,INT((SCOL()-18)/2)+17 ;
            TITLE "BCI Species Code List" ;
            NOFLOAT ;
            NOCLOSE ;
            SHADOW ;
            COLOR SCHEME 8
    ENDIF

    #REGION 1
    DEFINE POPUP sppopup ;
```

Program 3.1.1B. (continued)

```
        PROMPT FIELD SPPLIST->MNEM ;
        SCROLL ;
        MARGIN ;
        MARK "▶ "

    #REGION 1
    IF WVISIBLE("specscrn")
        ACTIVATE WINDOW specscrn SAME
    ELSE
        ACTIVATE WINDOW specscrn NOSHOW
    ENDIF
    @ 0,2 GET mDataAbbr ;
        PICTURE "@&N" ;
        POPUP sppopup ;
        SIZE 17,11 ;
        DEFAULT " " ;
        COLOR SCHEME 2

    IF NOT WVISIBLE("specscrn")
        ACTIVATE WINDOW specscrn
    ENDIF

    READ

    RELEASE WINDOW specscrn

    SELECT (m.currarea)

    RELEASE POPUPS sppopup

    #REGION 0
    IF m.talkstat = "ON"
        SET TALK ON
    ENDIF
    IF m.compstat = "ON"
        SET COMPATIBLE ON
    ENDIF

PROCEDURE loctag

    @8,15 TO 16,60
    @10,24 SAY "Start with tag number:"
    mTMP=.T.
    DO WHILE mTMP
        @12,34 GET mTAG PICTURE '999999' COLOR G/B, R/W
        READ
        SET INDEX TO obstag
```

Program 3.1.1B.(continued)

```
        FIND &mTAG
        IF FOUND()
            m.saverecno=RECNO()
            mTMP=.F.
            SET INDEX TO OBSV
            GO m.saverecno
        ELSE
            WAIT WINDOW "...tag not found..."
        ENDIF
    ENDDO
    CLEAR
    RETURN
PROCEDURE Plus
    SET CARRY TO &field_string
    KEYBOARD '{CTRL+N}' + &key_string + '{ENTER}' + mTAGtmp + ;
            '{HOME}' + '{HOME}' + '{BACKSPACE}'
    mTAGtmp = PADL(LTRIM(STR(VAL(TAG)+1)), 6, '0')
    RETURN
PROCEDURE Minus
    SET CARRY TO &field_mult
    KEYBOARD '{CTRL+N}' + &key_string + '{ENTER}' + '{ENTER}'
    RETURN
```

Box 3.1.1. The main data entry program MAIN.PRG.

The data entry program for the main data sheet is given as Program 3.1.1A, with procedures as Program 3.1.1B. The former is in a file called MAIN.PRG and the latter in PROC.PRG. PROC.PRG also includes procedures for the other two data entry programs (Programs 3.1.2, 3.1.3). In the printed version of these programs, shown as Program 3.1.1A and 3.1.1B, there are line numbers along the left margin. These are there only for reference in the following description, and the program will not run if they are included.

To use MAIN.PRG a database called MAINDATA.DBF must be created before data entry begins. It is created once—subsequent data are added to this same file. If data are to be double-entered, a second database must be created and given the same name, MAINDATA.DBF; it must be stored in a different subdirectory or on a different computer. Ideally, each data clerk has one computer, building his or her own version of MAINDATA.DBF.

MAINDATA.DBF must have the following fields: 1) *NAME* (for the field worker's or team's designation), 2) *DATE* (for the census date of the quadrat), 3) *CHQ* (initials of supervisor who checked the quadrat), 4) *DATE_CHQ* (date of checking), 5)

DATAENT1 (initials of first clerk entering data), 6) *DATE_ENT1* (date of first entry), 7) *DATAENT2* (initials of second clerk entering data), 8) *DATE_ENT2* (date of second entry), 9) *Q20* (quadrat number), 10) *P5* (subquadrat number), 11) *TAG* (tree's tag number), 12) *MNEM* (species mnemonic), 13) *DBH* (tree's dbh), and 14) *CODE* (tree's codes). All are FoxPro character data types except the four dates, which are date types, and *DBH*, which is a numeric variable. Fields 1-9 are only entered once per quadrat. Fields 10-13 must be entered for each record.

The program also requires a species database called SPPLIST.DBF with a field *MNEM* that contains species codes. As data are entered, the mnemonic of each record is screened against this list.

At the start of the program, at lines 1-5, are a series of simple commands that will appear near the beginning of most programs presented here. *CLEAR* and *CLEAR ALL* clear the screen and remove any databases that might be open. *SET CONFIRM ON* forces the clerk to move between fields with the enter key, while *SET BELL OFF* and *SET TALK OFF* turn off FoxPro features that provide information to the user about the status of the cursor and the database, but would only serve to slow down entry. Line 7 indicates where the subroutines are located, and lines 8-11 are variable initializations.

When executed, the program first prompts the user for his or her initials and the date (subroutine *prelim*, line 13 in MAIN.PRG, lines 1-10 in PROC.PRG). These are stored in the fields *DATAENT1* and *DATE_ENT1* (at line 41). When a second data clerk enters the data, line 41 must be changed to list *DATAENT2* and *DATE_ENT2*.

The browse window then opens, appearing like a standard FoxPro blank database. The fields listed in lines 44 to 59 are the only fields that will appear on the screen. The tab key, enter key, or arrow keys can be used to move between fields, just as when editing any FoxPro database. But the program has two features to facilitate rapid and accurate data entry. First, data screening is built in, and second, fields that do not change often are automatically copied when a new record is added.

Screening is based on the FoxPro commands shown in lines 44-65. First, the *Q20* must be legal—the first two digits must fall between 00 and 49 and the last two between 00 and 24. The *:V* on line 44 is a FoxPro command for checking validity, and will automatically produce on the screen a window announcing an invalid entry if *Q20* does not pass the tests in lines 44 and 45. *DATE* and *DATE_CHQ* are checked with the *:B* command, which compares them against a range of values. This command also produces a window listing the allowed range when an incorrect value is entered. In addition, each of the digits in *P5* must be between 1 and 4, *DBH* has to fall between −1 and 4000 mm, and only certain characters are allowed in the *CODE* field (B, D, M, R, N, T, C, S, L, Q, or V). Lines 44 to 65 can be adjusted to allow different legal values for all these fields.

In addition, the species codes are checked against the codes listed in SPPLIST.DBF, using the procedure *spLookUp*, which is called at line 56. The procedure (lines 13-28, Program 3.1.1B) scans the species list, and if the name entered is not there, a series of question marks appear in the *MNEM* field, and a popup window with the species list appears on the screen (created by the procedure *speclist*, lines 31-99, Program 3.1.1B). The window allows a name to be chosen. If the record requires a name which is not in the species list database, it would have to be edited later, after the new name is added to the species list.

Copying fields into the next blank record relies on the FoxPro *SET CARRY TO* command. Lines 23-30 arrange for fields 1-9 to be copied to the next record by assigning to the variable *field_string* the names of the first nine fields. In addition, the subroutine *Plus* configures the plus key to skip to the *P5* field (the 10th field) in the next record and to fill in the next highest tag number (lines 32-34 in MAIN.PRG, lines 124-130 in PROC.PRG). The clerk can then avoid tabbing across the records that were filled in automatically. (The plus key can be reset to its normal function by entering *ON KEY* on the FoxPro command line.) The copied fields can still be changed, though, simply by moving the cursor back. For example, when starting a new quadrat, *Q20* initially gets the value of the previous quadrat, but can then be edited.

Program 3.1.2. Data entry program for original trees, those tagged in a prior census (recensus only).

Program ORIG.PRG

```
CLEAR
CLEAR ALL

SET CONFIRM ON
SET BELL OFF
SET TALK OFF

SET PROCEDURE TO PROC

mTAG='    '
mD=ctod(' / / ')
mN=' '
mF=ctod(' / / ')

DO prelim

SELECT 0
USE OBSV1

DO loctag

REPLACE DATAENT1 with mN, DATE_ENT1 with mF

start_fld=7
key_string=IIF(start_fld>1, REPLICATE('+chr(9)', start_fld-1), '')
PUSH KEY
ON KEY LABEL + KEYBOARD '{ENTER}' + '{DNARROW}' + &key_string

BROWSE LAST NOWAIT FIELDS ;
  Q20       :R, ;
```

Program 3.1.2. (continued)

```
31    P5      :R, ;
32    TAG     :R, ;
33    MNEM    :R, ;
34    CODE    :R, ;
35    NAME    :P='!!!' ;
36            :H='INIT', ;
37    Date    :B=CTOD('01/01/95'),CTOD('12/31/95') ;
38            :H='DATE', ;
39    Dbh     :B=-1,4000 ;
40            :H='DIAMETER', ;
41    Code    :P='!!!!!' ;
42            :H='CODES' ;
43            :V=SUBSTR(code,1,1)$[ BDMRNTCSLQVE.] AND ;
44               SUBSTR(code,2,1)$[ BDMRNTCSLQVE.] AND ;
45               SUBSTR(code,3,1)$[ BDMRNTCSLQVE.] AND ;
46               SUBSTR(code,4,1)$[ BDMRNTCSLQVE.] AND ;
47               SUBSTR(code,5,1)$[ BDMRNTCSLQVE.]
```

Box 3.1.2. Data entry program for original trees, ORIG.PRG.

The data entry program used for original trees—those tagged in an earlier census (and thus only relevant in a recensus)—is shown as Program 3.1.2; it also uses the procedures listed in Program 3.1.1B. Unlike the main program (3.1.1A), this one requires that data from the prior census be stored in a FoxPro database. The database file is called OBSV1, and it must have the fields: 1) *NAME* (for the field team's designation), 2) *DATE* (for the census date of the quadrat), 3) *DATAENT1* (initials of first clerk entering data), 4) *DATE_ENT1* (date of first entry), 5) *DATAENT2* (initials of second clerk entering data), 6) *DATE_ENT2* (date of second entry), 7) *Q20* (quadrat number), 8) *P5* (subquadrat number), 9) *TAG* (tree's tag number), 10) *MNEM* (species mnemonic), 11) *OLDCODE* (codes from prior census, 12) *DBH* (tree's current dbh), and 13) *CODE* (tree's current codes). All are FoxPro character data types except for the three dates, which are date types, and *DBH*, which is a numeric variable. Fields 1-6 are only entered once per quadrat. Fields 7-11 are data from the previous census, and are already stored in the file OBSV1; they are not entered. Just fields 12 and 13 require data entry for each record.

The program also requires two index files: OBSTAG.IDX, an index on *TAG* for OBSV1, and OBSV.IDX, an index on *Q20+P5+TAG*. Indexing on tag means that each record is given an index based on its position in a numerical sequence of all tag numbers. This allows the program to immediately locate any tag number. Indexing on quadrat plus subquadrat plus tag means an ordering based on quadrat, within quadrat based on subquadrat, and within subquadrat based on tag. Indices are quickly created in the database with a single FoxPro command.

When executed, the program first prompts the user for his or her initials and the current date, using the same procedure *prelim* that MAIN.DBF uses (called at line 15). Next, the program asks the user for the tag number where data entry will start. This can be anywhere in the file—the program uses the *TAG* index to find this record, using the procedure *loctag* (called at line 20 in ORIG.PRG; lines 101-121 PROC.PRG). A browse screen then opens (line 31), showing the record of the OBSV1 database that *loctag* found. Fields from the prior census are already filled in, and these are designated read-only with the command *:R* (lines 30-34). The *NAME*, *DATE*, *DBH*, and *CODE* fields are blank, and data are entered. After finishing one record, the clerk uses the plus key to to skip to the *DBH* field for the next record (configured at lines 24-27).

Data are screened for validity at lines 35-47 exactly as in MAIN.PRG.

Program 3.1.3. Data entry program for multiple stems.

Program MULT.PRG

```
CLEAR
CLEAR ALL
SET CONFIRM ON
SET BELL OFF
SET TALK OFF

SET PROCEDURE TO PROC

PUBLIC key_string, field_string, field_mult

mN='   '
mF=ctod('  /  /  ')

DO prelim

SELE 0
USE MULT1
GOTO BOTTOM

start_fld=10
field_string=FIELD(1)
N=1
DO WHILE N < start_fld
   field_string=field_string+','+FIELD(N)
   N=N+1
ENDDO

field_mult=FIELD(1)
N=1
   DO WHILE N < start_fld+2
   field_mult=field_mult+','+FIELD(N)
N=N+1
ENDDO
```

Program 3.1.3. (continued)

```
34
35   key_string=IIF(start_fld>1, replicate('+chr(9)', start_fld+3), '')
36   PUSH KEY
37   ON KEY LABEL + DO Plus
38   ON KEY LABEL - DO Minus
39
40   SELECT MULT1
41   IF RECCOUNT()=0
42      APPEND BLANK
43   ENDIF
44
45   REPLACE DATAENT1 with mN,DATE_ENT1 with mF
46
47   BROWSE LAST NOWAIT FIELDS ;
48     Q20       :V=VAL(SUBSTR(q20,1,2))>=0 AND VAL(SUBSTR(q20,3,2))>=0 and ;
49                   VAL(SUBSTR(q20,1,2))<=49 AND VAL(SUBSTR(q20,3,2))<=24,;
50     NAME      :P='!!!' ;
51               :H='INIT', ;
52     Date      :B=CTOD('01/01/95'),CTOD('12/31/95') ;
53               :H='DATE', ;
54     Chq       :P='!!!' , ;
55     Date_chq  :B=CTOD('01/01/95'),CTOD('12/31/95') ;
56               :H='Dt_cheq', ;
57     P5        :V=p5>='11' AND substr(p5,1,1)<='4' AND substr(p5,2,1)<='4', ;
58     Tag       :P='999999', ;
59     Dbh       :B=-1,4000 ;
60               :H='Diam'
```

Box 3.1.3. Data entry program for multiple stems, MULT.PRG.

The data entry program for entering dbhs of multiple-stemmed trees is listed as Program 3.1.3. The file is called MULT.PRG, and it also uses procedures in PROC.PRG (Program 3.1.1B).

To use MULT.PRG, a database called MULT1 must be created before data entry begins. It is only created once; subsequent data are added to this file. Again, if data are double-entered, a second copy must be created as well. MULT1 must have the following fields: 1) *NAME* (for the field team's designation), 2) *DATE* (for the census date of the quadrat), 3) *CHQ* (initials of supervisor who checked the quadrat), 4) *DATE_CHQ* (date of checking), 5) *DATAENT1* (initials of first clerk entering data), 6) *DATE_ENT1* (date of first entry), 7) *DATAENT2* (initials of second clerk entering data), 8) *DATE_ENT2* (date of second entry), 9) *Q20* (quadrat number), 10) *P5* (subquadrat number), 11) *TAG* (tree's tag number), and 12) *DBH* (tree's dbh).

MULT.PRG works just like ORIG.PRG and MAIN.PRG. After entering the user's initials and the date, the program opens the browse window (line 49). Multiple dbhs from a single tree are entered on multiple lines, each having the same tag number. Most fields are carried over from record to record, exactly as in the program MAIN.PRG. MULT.PRG uses the *Plus* subroutine (line 37) to skip to the *P5* field of the next record; since the subroutine automatically augments the tag number by one, the tag field must be edited (since multiple stems are seldom found on successive stems). In addition, MULT.PRG uses a new subroutine, *Minus* (line 38), which formats the minus key to skip to the *DBH* field of the next record, skipping *P5* and *TAG* as well. The *Minus* subroutine is found at line 132-136, Program 3.1.1B. Data screening in MULT.PRG is done as in ORIG.PRG and MAIN.PRG (lines 50-62).

Program 3.1.4. Comparison program for detecting discrepancies between two data files.

Program COMPARE.PRG

```
SET TALK OFF
SET SCOREBOARD OFF
SET SAFETY OFF
CLEAR ALL
CLEAR

@12,12 SAY "Verification program is running....Please Wait"

mSTOP=0
prevREC=0


SELECT 1
USE FILE1
mNumFLD=FCOUNT()
DIMENSION aFIELDN(mNumFLD)
N=1
DO WHILE N<=mNumFLD
   aFIELDN(N)=FIELD(N)
   N=N+1
ENDDO

SELECT 2
USE FILE2
SELECT 3
USE DATACOMP
SELECT 2

DO WHILE NOT EOF()

   IF mSTOP>=5
      @ 14,22 SAY "Program ended - check if one of files skipped a record"
```

Program 3.1.4. (continued)

```
        EXIT
    ENDIF
    mAPPBLNK=.T.
    N=1

DO WHILE N<=mNumFLD
    IF &aFIELDN(N)<>a->&aFIELDN(N)

        gorec=RECNO()
        mTMP2=&aFIELDN(N)
        mTMP1=a->&aFIELDN(N)

        SELECT 3
        IF mAPPBLNK
            APPEND BLANK
            REPLACE RECORDNO WITH gorec,&aFIELDN(N) with mTMP1
            APPEND BLANK
            REPLACE RECORDNO WITH gorec,&aFIELDN(N) with mTMP2
            mAPPBLNK=.F.
            IF gorec=prevREC+1
                mSTOP=mSTOP+1
            ELSE
                mSTOP=1
            ENDIF
            prevREC=gorec
        ELSE
            SKIP-1
            REPLACE &aFIELDN(N) WITH mTMP1
            SKIP
            REPLACE &aFIELDN(N) WITH mTMP2
        ENDIF

    ENDIF

    SELECT 2
    N=N+1

ENDDO

SELECT 1
IF NOT EOF()
    SKIP
ENDIF
SELECT 2
SKIP

ENDDO
```

Program 3.1.4. (continued)

```
80
81	CLEAR ALL
82	SET TALK ON
83	SET SCOREBOARD ON
84	SET SAFETY ON
```

Box. 3.1.4. Program for comparing double-entered data files, COMPARE.PRG.

Program 3.1.4 opens two FoxPro data files which are ostensibly identical and locates any discrepancies between the two. Discrepancies indicate data entry errors. A third FoxPro file is created which lists all the mismatches.

The two files to be opened are FILE1 and FILE2. These files can include any number of variables, but they must match, having the same number of variables with the same names. A third file must be created in advance, called DATACOMP, with the same set of variables but no records. DATACOMP also needs one additional integer variable called *RECORDNO* to store the record number of each discrepancy.

Lines 13-26 open the three data files in work spaces 1, 2, and 3. The number of variables (fields) is counted (line 15), and the variable names are stored in an array (lines 18-21). At line 29, a long loop reads through successive records, and at line 38, a sub-loop reads through all fields of each record. Lines 39-65 compare the values for each variable from the two original files: because 2 is selected (line 27), *&aFIELDN(N)* refers to the value in the file FILE2 and *a->&aFIELDN(N)* refers to work space 1, hence file FILE1. The *&* means that the values of the variables are compared. If the two values differ, the current record number (stored in *gorec*) is copied into the variable *RECORDNO* of the new file, and the values for the two versions of the variable are copied into the same variable name—on two consecutive lines—in the output file, DATACOMP. The reason for the *IF* loop at lines 46-63 is that a blank record must be appended to the output file when the first error is found on a given line; this sets *mAPPBLNK* to.*F*.. Subsequent errors on the same line do not require the blank record, and the program jumps to line 58. Thus, all discrepanices found on one line in the two original files are written to two lines in the output file, one for each file.

The program also counts consecutive lines with errors (lines 52-56). The variable *mSTOP* is augmented by 1 whenever two consecutive lines have discrepancies, and the entire program exits when *mSTOP* reaches 5 (lines 31-34). In nearly all cases where there are five consecutive lines with errors, the cause is a missing line in one of the original files. From then on, all lines will not match, and every single line will be written to the output file. Use of *mSTOP* prevents this but still reveals the missing line. Once the missing line is added, the comparison program must be run again.

Program 3.1.5. Program for reading individual map files from each quadrat into a large FoxPro database for the whole plot.

Program APPMAP.PRG

```
CLEAR ALL
CLEAR
SET SCOREBOARD OFF
SET TALK OFF
SET SAFETY OFF

@12,25 SAY "<<Appending Maps...Please Wait>>"

USE TREELOC

mCol=0
DO WHILE mCol<50
   mRow=0
   DO WHILE mRow<25
      COUNT TO numRecs
      mQ20=PADL(ALLTRIM(STR(mCol,2)),2,'0')+;
            PADL(ALLTRIM(STR(mRow,2)),2,'0')
      @14,30 SAY "Working on Quadrat "+mQ20
      i=1
      DO WHILE i<=4
         mQuad='Q' + mQ20 + '.' + STR(i,1)
         IF FILE(mQuad)
            APPEND FROM &mQuad DELIMITED
            GO numRecs+1
            REPLACE Q20 WITH mQ20, Q10 WITH i WHILE NOT EOF()
         ENDIF
         i=i+1
      ENDDO
      mRow=mRow+1
   ENDDO
   mCol=mCol+1
ENDDO

REPLACE ALL GX WITH 20*INT(VAL(Q20)/100)+LX,;
      GY WITH 20*VAL(SUBSTR(Q20,3,2))+LY FOR Q10=1

REPLACE ALL GX WITH 20*INT(VAL(Q20)/100)+LX,;
      GY WITH 20*VAL(SUBSTR(Q20,3,2))+LY+10 FOR Q10=2

REPLACE ALL GX WITH 20*INT(VAL(Q20)/100)+LX+10,;
      GY WITH 20*VAL(SUBSTR(Q20,3,2))+LY FOR Q10=3

REPLACE ALL GX WITH 20*INT(VAL(Q20)/100)+LX+10,;
```

Program 3.1.5. (continued)

```
44        GY WITH 20*VAL(SUBSTR(Q20,3,2))+LY+10 FOR Q10=4
45
46    SET SCOREBOARD ON
47    SET TALK ON
48    SET SAFETY ON
49    CLEAR ALL
50    CLEAR
```

Box 3.1.5. Program for appending map files into one large map database, APPMAP.PRG.

Program 3.1.5 opens all 5000 map files in order and combines them into a single FoxPro database. Each individual file was created by the digitizing tablet, and holds coordinates for each plant within a single field map—a 10 x 10 m subregion. These are the local coordinates, *LX* and *LY*, perpendicular distances from the lower left corner of each subregion (so always < 10). Files were given names such as Q3513.1, where 3513 is the quadrat (20 x 20 m) number, and 1 is the 10x10 m subregion within. They are saved as text files delimited with commas.

Before running Program 3.1.5, an empty FoxPro database called TREELOC must be created, and must include the following variables: *TAG, LX, LY, Q20, Q10, GX, GY*. It will be filled in by the program. The first three variables will be copied directly out of the map file, the next two are created using the name of the file, since the name carries the quadrat number. The last two are global coordinates, perpendicular distances from the lower left corner of the entire plot.

Program 3.1.5 runs through a loop of column numbers from 0 to 49 (line 12), a loop of row numbers from 0 to 24 (line 14), and a loop of 10x10 m subregions from 1 to 4 (line 20). At lines 16 and 21, these numbers are converted into a file name, first Q0000.1, then Q0000.2, all the way up to Q4924.4. Each of the map files with the corresponding names is opened and its contents are appended into TREELOC at line 23. The quadrat and *Q10* variables are copied for each record (line 25).

After all the files have been appended, global coordinates are calculated for each record (lines 34-44). The calculation is illustrated in Figure 3.1.2.

The program is appropriate for a 50 ha plot with 50 columns and 25 rows. For a plot with a different shape, the column and row loops would have to be adjusted. Also, in a recensus, maps cover entire 20 x 20 m quadrats, so 10x10 m subregions need not be considered. The program would have to be shortened, removing the subregion extension (lines 19-21), and all *GX* and *GY* coordinates are calculated as in lines 34-35 (the condition *FOR Q10=1* and lines 37-44 would be removed).

Program 3.1.6. Program for detecting duplicate tags.

Program DUPTAG.PRG

```
SET STATUS OFF
SET TALK OFF
CLEAR
CLEAR ALL

filen='        '
@8,5 SAY “Enter the filename:”
@8,26 GET filen PICTURE “@!”
READ
filen2=RTRIM(filen)+'.dbf'
n = 8

DO WHILE NOT FILE(filen2)
    @n,4 SAY “Filename is not correct - Enter correct filename:”
    @n,54 GET filen PICTURE “@!”
    READ
    filen2=RTRIM(filen)+'.dbf'
    n=n+1
ENDDO

mTAG=””
@15,2 SAY;
        “<<Program Checking for Duplicate Tags Running.....Please Wait>>”

SELECT 1
USE &filen
INDEX ON TAG TO &filen
SELECT 2
USE DUPLERRS
SELECT 1
mTAG=TAG
DO WHILE NOT EOF()
    SKIP
    IF TAG=mTAG
        SKIP-1
        COUNT TO mDUPL WHILE TAG=mTAG
        SELECT 2
        APPEND BLANK
        REPLACE TAG WITH mTAG, NoDUPL with mDUPL
        SELECT 1
    ENDIF
    mTAG=TAG
```

Program 3.1.6. (continued)

```
43    ENDDO
44
45    CLOSE ALL
46    idxfile = filen + '.IDX'
47    ERASE &idxfile
48    SET TALK ON
49    SET STATUS ON
50    @18,10 say "....DONE!!!"
```

Box 3.1.6. Program for locating duplicate tags, DUPTAG.PRG.

Program 3.1.6 is a FoxPro program that screens any FoxPro data file for duplicate records in the tag field. It is stored as the file DUPLERRS.DBF. Unlike prior FoxPro programs, this lets the user enter the file name (lines 6-9), and requests a different name if the file does not exist (lines 13-19). The program works by indexing on the tag field, which creates an index file with the same name as the main file, but with extension IDX (note that this file is erased when the program is done; line 49). The indexing means that duplicate tag numbers will have consecutive indices. Thus, the current tag number is stored (lines 31 and 44), then compared to the next tag number (line 34). If they match, the record is appended to the new database called DUPLERRS (opened at lines 28-29).

The file DUPLERRS must be created in advance with two fields, for tag and number of duplicates (NODUPL). When the program finishes, this database has a list of all tags that appeared in duplicate.

CHAPTER 3.2

Constructing a Database for the First Census

Carefully organizing the large plot data in the computer can greatly facilitate analyses. For demographic work, the key features of a plant are its species, location, size (dbh), the date at which its size was measured, whether its dbh in one census was recorded at the same spot as in the previous census (so growth can be estimated), and whether the plant is alive or dead. In this and the next chapter, I describe how to create a database that allows these features to be readily accessed. Up to this point, as described in chapter 3.1, the census data have been entered in the computer, and a main database, coordinates database, and species names all merged into one large file. Two smaller files contain dbhs of multiple-stem plants and big trees. Corrections in the database have been completed. Table 3.2.1 reviews the data fields present in the three files. I refer to these files as the provisional database, and they will be merged and edited to create the final database.

In this chapter, I consider only a final database for the first census. This is relatively simple, as there are no dead trees yet nor concern about changes in the stem measurement point (complexities I cover in the next chapter). Nearly all the information necessary for analyses after one census is already in the three computer databases created by the programs given in chapter 3.1 (Table 3.2.1). Some of this information goes straight into the final database. Other fields are omitted and stored separately, since they are not necessary for data analysis.

Data Fields in the Final Database

The Main Elements

The quadrat, subquadrat, tag, dbh, and codes were all recorded on the data sheets in the field by the mapping crew, and these were entered directly in the computer by data clerks. They remain unchanged in the final database. The taxonomy team assigned species codes in their taxonomic database (chapter 2.3), and this was merged with the main database after data entry was complete (chapter 3.1).

Census Date

The date on which trees were mapped and measured is an important piece of information, particularly for growth and mortality calculations which become possible with the second census. The date already appears for every individual tree: dates were originally entered once per quadrat by clerks, but the data-entry

Tropical Forest Census Plots: Methods and Results from Barro Colorado Island, Panama and a Comparison with Other Plots, by Richard Condit.

Table 3.2.1. Data fields in the three provisional databases from a first plot census. These were created by the programs given in chapter 3.1 on data entry.

A) Main file

data field	contents
person collecting data (initials)	2 characters
date censused	date
person checking data (initials)	2 characters
date checked	date
1st person entering data (initials)	2 characters
1st date entered	date
2nd person entering data (initials)	2 characters
2nd date entered	date
quadrat number	4-digit integer
subquadrat number	2-digit integer
tag number	6-digit integer
species name	6-letter code
x coordinate (m)	floating point number, 1 digit after decimal
y coordinate (m)	floating point number, 1 digit after decimal
dbh (mm)	4-digit integer
codes	up to 6 characters in a string

B) Multiple-stem file

data field	contents
person collecting data (initials)	2 characters
date censused	date
person checking data (initials)	2 characters
date checked	date
1st person entering data (initials)	2 characters
1st date entered	date
2nd person entering data (initials)	2 characters
2nd date entered	date
quadrat number	4-digit integer
subquadrat number	2-digit integer
tag number	6-digit integer
dbh (mm)	4-digit integer

C) Big-tree file

data field	contents
person collecting data (initials)	2 characters
date censused	date
person checking data (initials)	2 characters
date checked	date
1st person entering data (initials)	2 characters
1st date entered	date
2nd person entering data (initials)	2 characters
2nd date entered	date
quadrat number	4-digit integer
subquadrat number	2-digit integer
tag number	6-digit integer
dbh (mm)	4-digit integer
height of POM (m)	floating point number, 2 digits after decimal

program automatically copied it for every record. However, later corrections may have required changing some individual dates. Thus, each tree requires its own date, and this field must remain in the final database.

Multiple-Stems and Big Trees

Multiple-stemmed plants and big trees still do not have anything in their dbh column. To fill this in, the multiple-stem database must be consulted and the largest dbh copied and inserted in the main database. In addition, the final database stores a new variable—the number of stems on each plant. All records without the code 'M' have 1 stem; for the rest, the number of stems in the multiple-stem database must be counted. Program 3.2.1 shows a FoxPro routine which opens the main and the multiple-stem databases, extracts the largest dbh from the latter, counts the number of stems, and inserts this information in the main database. The multiple-stem database is unchanged by these procedures, and is stored for future use (it is the only place where dbhs of secondary stems are kept and thus will be needed for some analyses).

A similar procedure must be carried out for big trees, with the dbh taken from the big trees file. Generally, during a first census, the big trees have only a single dbh taken above the buttresses, so no decision needs to be made about which dbh to use. The multiple-stem program (program 3.2.1) can be used for big trees with little modification (Box 3.2.1).

Height of Measure

The height of the POM (point-of-measure) is not necessary for analyses, but will be necessary on field sheets in the future censuses. In the big-tree database, the height of the POM was entered in a separate field (chapter 3.1), and is stored there but need not be copied into the main database. In smaller plants where the POM was not at 1.3 m, the height of measure was recorded in the codes column. It was entered in the main database with the other codes and remains in the final database.

New Data Fields

Three new data fields go in the final database. One is the number of stems, which was filled in from the multiple-stem database by Program 3.2.1 (see Box 3.2.1). The two remaining new fields are not necessary for a first-census database, but should be inserted at this point so that a first census database matches the database from a recensus; they will be necessary when two censuses are available. The advantage of adding them now is that it allows analytic programs for handling a multiple-census database to also handle a single-census database without modification.

One of the new fields is the *POM* number, which always takes the value 1 for the first census (so every single record gets a number 1). How *POM* becomes important in future censuses will be described in the next chapter. The other new field is the status code, a single character which indicates if a plant is alive or dead. In the first census at most large plots, all plants are alive, so each gets a code 'A'. (If dead trees were mapped during the first census, as was done in the Mudumalai plot in India, they are given a status 'D'. After a second census, the status code can take two other values, described in chapter 3.3.)

In the final database, the original codes that were recorded on the field sheets are referred to as general codes, whereas the newly created status field is referred to as a status code (Table 3.2.2).

Data Fields Removed from the Final Database

Seven of the data fields in the provisional database do not have to be kept in the final database; they are not necessary for analyses. These are the names of personnel and dates of field checks and data entry (Table 3.2.1). This information can be stored just once for each quadrat (not for every plant), so a new quadrat database should be created to hold this information. This is easily done in FoxPro with the *UNIQUE* command, which extracts only those records unique for any indexed field, in this case, *Q20*, and these can be copied to a new database. The subquadrat, tag, dbh, species name, and code fields must be removed from this new database, while the date and name fields are kept. The new database has 1250 records for 1250 quadrats (in 50 hectares). The quadrat data in it are largely for logistical and administrative purposes and are seldom consulted again (chapter 3.1). The seven administrative fields can now be removed from the main database.

Splitting the Database in Two

The final database is now split into two on the following logic. The species name, quadrat, subquadrat, and coordinates are permanent attributes of each plant, and are stored in one database. They will never change through future censuses. The dbh, POM, stem number, date, and codes will change in the next census, and are stored in a second database. Tag number must appear in both to identify the records. This division into permanent and census databases anticipates future censuses, as I explain in the next chapter. FoxPro easily splits a database in two by copying the desired fields into new files.

Table 3.2.2 lists the data fields that remain in the final database. The multiple-stem and big-tree databases are unchanged from Table 3.1.1. I include the quadrat database for completeness, even though it is not used in analyses.

Table 3.2.2. Final first census database, ready for analysis.

A) Permanent database

data field	contents
tag number	6-digit integer
species name	four-letter code
quadrat number	four characters
subquadrat number	two characters
x coordinate (m)	floating point number, 1 digit after decimal
y coordinate (m)	floating point number, 1 digit after decimal

B) Census database

data field	contents
tag number	6-digit integer
dbh (mm)	4-digit integer
POM number	2-digit integer
stem number	3-digit integer
status code	1 character
census date	date
general codes	up to 6 characters in a string

C) Multiple-stem database

data field	contents
tag number	6-digit integer
dbh (mm)	4-digit integer

D) Big-tree database

data field	contents
tag number	6-digit integer
dbh (mm)	4-digit integer
height of POM (m)	floating point number, 2 digits after decimal

E) Quadrat database

data field	contents
quadrat number	4-digit integer
person collecting data (initials)	2 characters
date censused	date
person checking data (initials)	2 characters
date checked	date
1st person entering data (initials)	2 characters
1st date entered	date
2nd person entering data (initials)	2 characters
2nd date entered	date

At BCI, the final database is a FoxPro version 2.5 file which holds all records from the entire census, sorted by tag, divided into the files listed in Table 3.2.2. In addition, we create ASCII text versions of the main database, and these are organized slightly differently. Instead of all records together, each species gets a single file, and the name of the file is the species mnemonic. The species name is removed from each file, since it would be redundant. Separate species files allow rapid analyses of individual species. One entire set of species files for the permanent data is stored in one file folder on the computer, and another set for census data kept in a different folder (this is diagrammed in chapter 3.3). Thus after one census, there are twice as many text files as there are species. Program 3.2.2 shows how FoxPro can read the database and write a set of text files for each species.

In addition, multiple-stem and big-tree databases are transcribed into text. They are not divided into species files, but instead each is stored as a single, large file.

The purpose of text files is 2-fold. First, text files can be readily loaded into alternate database systems, according to the preferences or needs of individual users. Second, they allow the database to be accessed by other mainstream programming languages such as C/C++, FORTRAN, Pascal, or Basic, which are more efficient for complex analytical needs.

Database Correction

Routine Screening

I recommended a few routine procedures for verifying error rates and weeding out errors (chapter 2.2). One was to randomly re-measure all trees in selected quadrats. This requires no new database methods: the second set of dbhs can be entered in the computer using Program 3.1.1. When this secondary database is created, the dbhs can be compared to those from the original (linking the two databases through tag number).

Another simple screening procedure is to re-identify randomly selected individuals. One way to do so is to randomly choose entire quadrats, just as for the dbh screen described above. A second way, which samples rare species more thoroughly, is to randomly extract and re-identify four individuals from each species (chapter 2.3). Program 3.2.3 is a FoxPro procedure for extracting random individuals from each species. The output must be sorted by quadrat and then printed onto field sheets for the taxonomists to use.

Ongoing Correction

These routine screening procedures locate a few errors, but mainly just indicate the overall error rate. Other errors remain, and corrections will be made on an ongoing basis. Visiting researchers who do field work in the plot at BCI take a certain pleasure in finding our mistakes. They always seemed quite pleased to feed me a slip of paper noting that tag number X that was supposed to be *Poulsenia armata* is actually something entirely different. Some seem to believe they are the first to detect a mistake in a data set with over 325,000 records. But there are many errors, and the inter-census interval provides some time to expunge as many as possible.

Errors are noted regularly, and a list of corrections should be assembled, however, they should only be entered into the main database every two years or so. When the corrections are made, records of all changes must be stored so that the differences between an old version and a new version of the database are documented. It is very important not to correct each error as soon as it is found, for then the database would be changed monthly, and it would be impossible to keep multiple copies of the database updated and uniform. Similarly, it is important that errors not be corrected separately in each copy of the database. One person (the database manager) should be responsible for compiling and correcting errors. When the corrections are made, the new version is distributed to all collaborating scientists, and the old versions discarded. If everyone makes his or her own corrections, everyone will have a different database. It is embarrassing, for example, to have two different reports from the same project listing different numbers of species or different numbers of individuals.

Backing Up!

When the final database is ready, and the last corrections made, it is of course crucial to make a duplicate copy and store it safely. Zip drives (Iomega) are convenient and have enough storage space for the entire database.

Field Sheets for the Next Census

Besides all the analytical work that is the raison-d'être of the project, the final database also is the basis for the next set of data sheets to be used five years later. These were described and illustrated in chapter 2.5. The mapping crew during the recensus needs a map for each quadrat, with each plant and its tag number appearing. Recruits are mapped on this sheet by the field teams. Program 3.2.3 illustrates FoxPro code that writes Sygraph plotting routines to draw maps from the main database (see Box 3.2.3).

In addition, sheets listing all tag numbers from the quadrat, sorted by subquadrat, become the data sheets for original trees during the recensus. The species name, dbh, and codes from the first census should appear on these sheets. Program 3.2.4 is FoxPro code for printing all this information in a compact form (see Box 3.2.4).

Big-tree field sheets are also made up with data from the prior census. This is easily done by copying to a file all individuals with code 'B' in the first census database plus any new ones with the code in the recensus, then sorting and printing this by quadrat. Blank fields should be inserted as well for recording the new dbh, codes, and height of measure.

The other field sheets (multiple stem, recruits) do not have old data on them. They are identical to the forms used during the first census.

Equipment, Time, and Labor

Computers

As discussed in the previous chapter, duplicating, reading, and manipulating databases with 200,000 or more records is very tedious without a Pentium-driven personal computer with a gigabyte hard drive. Frequent back-ups of these large files should be made as well, so a zip-drive should be used for readily copying the large data sets to disks.

Labor

Creating the final database from the preliminary files entered by the data clerks takes a database manager about two weeks of work.

Key Issues for Database Creation

- Main data fields—quadrat, subquadrat, tag, coordinates, species, dbh, and codes—are copied directly from the provisional database created by clerks
- The census date is stored for every record in the main database
- The largest dbh from multiple-stemmed trees is copied from the multiple-stem database to the main database
- The dbh of big trees must also be copied from the big-tree database into the main database
- Three new fields are created for every plant—the POM, status, and stem number
- The final database includes four FoxPro files: one with location and species name, a linked database for census data, and two more linked databases for multiple stems and big trees
- ASCII text files are created, one per species per census plus one per species for location, for exporting to other databases.

Program 3.2.1. Program to count the number of multiple stems, find the largest dbh, and insert these numbers in the main database for plants with multiple stems.

Program GETMULT.PRG

```
SET SCOREBOARD OFF
SET TALK OFF
CLEAR
CLEAR ALL

SELECT 2
USE MAINDATA
SET ORDER TO TAG TAG
SELECT 1
USE MULT1
INDEX ON TAG TO MULT1
GO TOP

DO WHILE NOT EOF()
    mtag=TAG
    mdbh=DBH
    mcount=1
    SKIP
    DO WHILE TAG=mtag
        mcount=mcount+1
        IF DBH>mdbh
            mdbh = DBH
        ENDIF
        SKIP
    ENDDO
    SELECT 2
    FIND &mtag
    IF FOUND()
        REPLACE STEM_NO WITH mcount,DBH WITH mdbh
        IF NOT 'M'$CODE
            REPLACE CODE WITH ALLTRIM(CODE)+'M'
        ENDIF
    ENDIF
    SELE 1
ENDDO

CLEAR ALL
SET SCOREBOARD ON
SET TALK ON
```

Box 3.2.1. Program GETMULT.PRG for reading data out of the multiple-stem file.

Program 3.2.1 is a FoxPro program which simultaneously opens the main database and the multiple-stem database and reads the largest dbh for each multiple-stemmed plant into the dbh field of the main database. It also counts the number of stems for each multiple-stemmed plant and stores this in the main database. The main database is the final database created by Program 3.1.1, carrying all information, including global map coordinates, for each record. Here it is named MAINDATA.DBF. The multiple-stem database was entered by Program 3.1.3 and holds all dbhs for plants with more than one stem; it is named MULT1.DBF. These files should already have been cross-checked to make sure that each record with a code 'M' in MAINDATA has a corresponding record in MULT1, and vice versa.

Before running the program, the MAINDATA database must be modified in two ways. First, a structural compound index on the *TAG* field must be created, using the *MODIFY STRUCTURE* option. This index file is automatically updated whenever MAINDATA is changed, and is thus useful for many future manipulations. Second, a new integer field *STEM_NO* must be created. This will contain the stem count for each record. Initially, it is filled in with 1 for every plant.

At lines 6-8, MAINDATA is opened in workspace 2, indexed on the tag field. The appearance of the word 'TAG' twice at line 8 is confusing, but it is just a coincidence: the first *TAG* is part of the FoxPro command, the second is the field being indexed. Thus, if there were an index file for the *Q20* field, the command to access it would be *SET ORDER TO TAG Q20*.

At lines 9-11, the MULT1 database is opened and also indexed on tag. Line 11 is an index command that does not require a prior index file; in fact, it creates a new index file, however, this index file is not automatically updated like the one created by *MODIFY STRUCTURE*. Whether one utilizes a permanent, updated index file (line 8), or creates a temporary one (line 11), depends on whether the index is likely to be used in the future. This program will work regardless, as long as both data files are indexed on tag. The name MULT1 is repeated at line 11, indicating that the index file will have the same name (a different name could have been used).

The indexed MULT1 database is opened at the first record, and the *DO-WHILE* loop starting at line 14 causes the program to read through the database record-by-record, until the end of the file (*EOF*) is reached. The loop ends at line 35 with *ENDDO*. Lines 15-17 store the first tag number, the first dbh, and set the variable which counts the number of stems (*mcount*) to 1. Then *SKIP* moves the program to the second record, and another *DO-WHILE* loop is started if the current tag (*TAG*) matches the previous tag (*mtag*). Because every plant in the multiple-stem file has at least two records, each with the same tag, the second record must match the first, so for the second record, the loop at lines 19-25 is executed. The stem count is increased by 1 (line 20), and the current dbh is stored in *mdbh* if it is larger than the first dbh (lines 21-23). *SKIP* (line 24) then moves on to the third record, and the loop will be repeated for each record with the same tag number.

When the tag number changes, the program continues with line 26, selecting MAINDATA. The *FIND* command (line 27) is simple but useful: it searches the entire database for the tag number *mtag*, and is almost instantaneous because of the indexing; this is why MAINDATA had to be indexed on tag. The & must precede the memory variable name when using the *FIND* command. When the tag number is located, *FOUND()* returns true, and lines 29-32 are executed. The variable *mcount* is copied into the *STEM_NO* field, replacing the 1 already there, and the maximum dbh *mdbh* is copied into the *DBH* field. The latter had been left blank during data entry. Finally, at lines 30-31, the code 'M' is added to the code field if it was not already there. The operation *'M'$CODE* is FoxPro's way of evaluating whether the *CODE* includes any letter 'M'. *ALLTRIM(CODE)* removes any blanks from the codes, then 'M' is added to *CODE.* (Checking for the code 'M' should be an unnecessary precaution, since the two files were already cross-checked.)

The program then returns to the MULT1 database (line 34) and back to the top of the *DO-WHILE* loop at line 14. A little bit of thinking should show that the *SKIP* statement is properly placed at line 18, and not at the top of the loop. When the entire MULT1 database is complete, lines 37-39 close the databases and return FoxPro to the default settings.

The same program can be used for reading the dbh out of the big-tree database into the main database. The database names at lines 10 and 11 must be changed, and the code 'M' at lines 30-31 must be changed to 'B', and lines 17 and 19-25 must be removed. The stem count variable *mcount* should be removed. If all big trees have just one dbh, as they should after a first census, the program will work simply by reading one line per tag and inserting the dbh in the main database.

Program 3.2.2. Program for writing text files for each species, with spaces between each variable.

Program MAKETEXT.PRG

```
SET SCOREBOARD OFF
SET TALK OFF

SET SAFETY OFF
CLEAR ALL

@12,20 SAY "<<Program Running.....Please Wait>>"

SELECT 1
USE MAINDATA
INDEX ON MNEM+TAG TO MNTAG
DO WHILE NOT EOF()
   mMn=MNEM

   @ 14,30 SAY "Working on Species: "+mMN

   COPY TO TMP WHILE MNEM=mMn
   SELECT 2
   USE MNLOC
   ZAP
   APPEND FROM TMP
   COPY TO \BCI\CENSUS\&mMn SDF FIELDS TAG,G1,GX,G2,GY
```

Program 3.2.2. (continued)

```
23
24        USE MNCENS1
25        ZAP
26        APPEND FROM TMP
27        REPLACE ALL JUL WITH DATE-CTOD('01/01/81')
28        COPY TO \BCI\CENSUS0\&mMn SDF FIELDS TAG,G1,DBH,G2,;
29                 STATUS,G3,POM,G4,STEM_NO,G5,JUL,G6,CODE
30
31        SELECT 1
32    ENDDO
33
34    CLEAR ALL
35    ERASE TMP.DBF
36    SET SCOREBOARD ON
37    SET TALK ON
38    SET SAFETY ON
```

Box 3.2.2. Program MAKETEXT.PRG for creating the ASCII text database.

Program 3.2.2 opens the main FoxPro database, MAINDATA.DBF, and writes data from each species to two separate text files. One file includes the tag number and global coordinates for each individual, the other includes the tag, dbh, POM, status code, stem number, and date from one census. Two temporary databases that match the structure of the text files must be created. One, MNLOC.DBF, includes the fields *TAG*, *GX*, and *GY*, plus two blank fields *G1* and *G2* alternating with these. The second, MNCENS1.DBF, includes fields for *TAG, DBH, STATUS, POM, STEM_NO, DATE,* and *GEN_CODE*, each one character wide, plus a new field called *JUL*. Alternating with these are fields named *G1* through *G6*, each one character wide, which remain blank. The blank fields simply serve to print a blank space between columns in the text file.

MAINDATA must first be indexed on species and tag (line 11), then all records from a single species are copied to a temporary database, TMP. The MNLOC database is opened and all previous records erased with *ZAP* (lines 19-20), then TMP is appended into MNLOC. The *COPY* command at line 22 creates a text file in directory \BCI\CENSUS named for the species copied (*&mMn*).

A similar procedure is repeated for the census data at lines 24-29. First, at line 27, the date is converted to the number of days since 1 January 1981 and stored in the field *JUL* (this allows date arithmetic with the text file). The data fields are copied to the text file alternating with the blank fields *G1* through *G6*. This text file is also named for the species being copied, but is placed in the directory \BCI\CENSUS0.

The *COPY* command assures that all records from one species are copied into one pair of files. The next species starts the process over again, *ZAPPING* the temporary files and writing new text files.

Program 3.2.3. Program for extracting four individuals per species from a FoxPro database.

Program RANDSPP.PRG

```
SET TALK OFF
SET SCOREBOARD OFF
SET SAFETY OFF
CLEAR ALL

randno=RAND(-1)
DIMENSION aRSEL(4,8),aRECNUM(4)

SELECT 2
USE RANDOM
SELECT 1
USE MAINDATA
INDEX ON MNEM+TAG TO MNTAG FOR DBH>=10

DO WHILE NOT EOF()
   mMN=MNEM
   @12,25 SAY "Working on species = "+mMN
   fstrec=RECNO()
   aRECNUM=0
   COUNT TO numpl WHILE MNEM=mMN

   IF numpl<=4
      GO fstrec
      COPY NEXT numpl TO ARRAY aRSEL FIELDS,TAG,MNEM,;
            Q20,P5,GX,GY,DBH,CODE
      SKIP
   ELSE
      i=1
      DO WHILE i<=4
         STORE INT(numpl*RAND()) TO randno
         IF randno<numpl
            GO fstrec
            SKIP randno
            IF ASCAN(aRECNUM,RECNO())=0
               aRSEL(i,1)=TAG
               aRSEL(i,2)=MNEM
               aRSEL(i,3)=Q20
               aRSEL(i,4)=P5
               aRSEL(i,5)=GX
               aRSEL(i,6)=GY
               aRSEL(i,7)=DBH
               aRSEL(i,8)=CODE
               aRECNUM(i) =RECNO()
```

Program 3.2.3. (continued)

```
44                    i=i+1
45                 ENDIF
46              ENDIF
47           ENDDO
48           SKIP numpl-randno
49        ENDIF
50
51        SELECT 2
52        APPEND FROM ARRAY aRSEL FOR MNEM=mMN
53        SELECT 1
54     ENDDO
55
56     CLOSE DATA
57     SET TALK ON
58     SET SCOREBOARD ON
59     SET SAFETY ON
```

Box 3.2.3. Program RANDSPP.PRG for the random identification check.

Program 3.2.3 opens the main database, MAINDATA.DBF (line 12), and extracts a list of four individuals per species, selected at random. For species with < 4 individuals in the database, all are extracted. The output is a list of individuals in a database RANDOM.DBF (line 10). To begin, MAINDATA must first be indexed on the species name (field *MNEM*) and *TAG* (line 13).

The program then reads through the records for one species, storing the record number at the start of that species in *fstrec* (line 18) and the number of records for that species in *numpl* (line 20). If there are four or fewer records, then all are copied into the array *aRSEL* (lines 22-25). The array was declared at line 7; it is two-dimensional, with each row storing data from one individual, and each column a different field (as listed in lines 24-25). But if there are more than four records, just four must be selected at random (lines 28-49). A random number less than the number of individuals is chosen (lines 31-32; the random sequence was initialized at line 6). Line 33 jumps this many records, and then stores that record in the *aRSEL* array (lines 35-43). First, though, line 34 checks whether that record was already used (in case the same random number is selected twice). The one-dimensional array *aRECNUM* stores a list of the record numbers already used, and line 34 checks whether the current record is already in the list—if it is not, lines 35-43 store the record and add the record number to *aRECNUM*. This is repeated for four records using the counter *i* (lines 28-29). Finally, the four elements of the array *aRSEL* are written to the new database (line 52).

Program 3.2.4. Program for creating field map for use in recensus. It is a FoxPro program which creates Systat programs to draw the maps. One Systat program is created for each quadrat.

Program PRINTMAP.PRG

```
SET TALK OFF
SET SCOREBOARD OFF
SET SAFETY OFF
CLEAR ALL
CLEAR

USE TEMPDATA
APPEND FROM MAINDATA

REPLACE ALL SUBTAG WITH SUBSTR(TAG,4,3)
REPLACE ALL SUBTAG with ALLTRIM(SUBTAG)+'R' FOR 'R'$CODE
REPLACE ALL SUBTAG with ALLTRIM(SUBTAG)+'D' FOR 'D'$CODE
REPLACE ALL LX WITH GX-20*(INT(GX/20)), LY WITH GY-20*(INT(GY/20))

GO TOP
DO WHILE NOT EOF()
   DO CASE
      CASE (DBH>=10 AND DBH<100) OR (DBH=-1 AND 'D'$CODE);
            OR (DBH=0 AND 'R'$CODE)
         REPLACE RELSIZ WITH 1
      CASE DBH>=100
         REPLACE RELSIZ WITH ROUND(DBH/100,0)
      OTHERWISE
         REPLACE RELSIZ WITH 0
   ENDCASE
   SKIP
ENDDO
CLEAR ALL

USE TEMPDATA
INDEX ON Q20 to Q20
DO WHILE NOT EOF()
   mQ20=Q20
   mF='Q'+mQ20
   COPY FIELDS SUBTAG, LX, LY, RELSIZ TO &mF FOR RELSIZ>0;
         WHILE Q20=mQ20


ENDDO

USE TEMPDATA
INDEX ON Q20 TO Q20 UNIQUE
```

Program 3.2.4. (continued)

```
SET TEXTMERGE ON TO SYSFIL.CMD NOSHOW
mF='Q'+Q20
\\SAVE <<mF>>
\IMPORT “<<mF>>.dbf”/TYPE DB3
SKIP
DO WHILE NOT EOF()
    mF='Q'+Q20
    TEXT
        SAVE <<mF>>
        IMPORT “<<mF>>.dbf”/TYPE=DB3
    ENDTEXT
    SKIP
ENDDO

CLEAR ALL

USE TEMPDATA INDEX Q20

DO WHILE NOT EOF()
    mQ20='Q'+Q20
    mF=mQ20+'.CMD'
    SET TEXTMERGE ON TO &mF NOSHOW
    \\BEGIN
    TEXT
        ORIGIN -5cm,21.5cm
        WRITE '\3<<SUBSTR(mQ20,2,2)>>,<<SUBSTR(mQ20,4,2)>>',
                /HEIGHT=8pt,WIDTH=8pt
        ORIGIN 13.5cm,21.5cm
        WRITE '\3<<SUBSTR(mQ20,2,2)>>,<<SUBSTR(mQ20,4,2)>>',
                /HEIGHT=8pt,WIDTH=8pt
        ORIGIN -1.5cm,21.5cm
        WRITE '\3FOREST DYNAMICS PROJECT',
                /HEIGHT=8pt,WIDTH=8pt
        ORIGIN -5cm,20.75cm
        WRITE '\h\wPlease write your name and date:'
        ORIGIN -5cm,20cm
        WRITE '\h\wNames:'
        ORIGIN 2cm,20cm
        WRITE '\h\w1. Data entry by:'
        ORIGIN 9cm,20cm
        WRITE '\h\w2. Data entry by:'
        ORIGIN -5cm,19.25cm
        WRITE '___________'
        ORIGIN -4.8cm,19.25cm
        WRITE '___________'
        ORIGIN 2cm,19.25cm
```

Program 3.2.4. (continued)

```
        WRITE '__________'
        ORIGIN 2.2cm,19.25cm
        WRITE '__________'
        ORIGIN 9cm,19.25cm
        WRITE '__________'
        ORIGIN 9.2cm,19.25cm
        WRITE '__________'
        ORIGIN -5cm,18.75cm
        WRITE '__________'
        ORIGIN -4.8cm,18.75cm
        WRITE '__________'
        ORIGIN 2cm,18.75cm
        WRITE '__________'
        ORIGIN 2.2cm,18.75cm
        WRITE '__________'
        ORIGIN 9cm,18.75cm
        WRITE '__________'
        ORIGIN 9.2cm,18.75cm
        WRITE '__________'
        ORIGIN -5cm,18.25cm
        WRITE '__________'
        ORIGIN -4.8cm,18.25cm
        WRITE '__________'
        ORIGIN 2cm,18.25cm
        WRITE '__________'
        ORIGIN 2.2cm,18.25cm
        WRITE '__________'
        ORIGIN 9cm,18.25cm
        WRITE '__________'
        ORIGIN 9.2cm,18.25cm
        WRITE '__________'

        USE <<mQ20>>
        CS=2.0
        SELECT LY<10

        ORIGIN=-5cm,-3.5cm
        PLOT LY*LX/xmin=0,XMAX=4.95,ymin=0,YMAX=4.95,
            XTICK=5,YTICK=5,SYMBOL=2,SIZE=RELSIZ,LABEL=SUBTAG$,
            XLABEL=' ',YLABEL='',SCALE=0,HEIGHT=5cm,WIDTH=5cm

        ORIGIN=0cm,-3.5cm
        PLOT LY*LX/xmin=5,XMAX=9.95,ymin=0,YMAX=4.95,
            XTICK=5,YTICK=5,SYMBOL=2,SIZE=RELSIZ,LABEL=SUBTAG$,
            XLABEL=' ',YLABEL='',SCALE=0,HEIGHT=5cm,WIDTH=5cm

```

Program 3.2.4. (continued)

```
ORIGIN=5cm,-3.5cm
PLOT LY*LX/xmin=10,XMAX=14.95,ymin=0,YMAX=4.95,
   XTICK=5,YTICK=5,SYMBOL=2,SIZE=RELSIZ,LABEL=SUBTAG$,
   XLABEL='',YLABEL='',SCALE=0,HEIGHT=5cm,WIDTH=5cm

ORIGIN=10cm,-3.5cm
PLOT LY*LX/xmin=15,XMAX=20,ymin=0,YMAX=4.95,
   XTICK=5,YTICK=5,SYMBOL=2,SIZE=RELSIZ,LABEL=SUBTAG$,
   XLABEL='',YLABEL='',SCALE=0,HEIGHT=5cm,WIDTH=5cm

ORIGIN=-5cm,1.5cm
PLOT LY*LX/xmin=0,XMAX=4.95,ymin=5,YMAX=9.95,
   XTICK=5,YTICK=5,SYMBOL=2,SIZE=RELSIZ,LABEL=SUBTAG$,
   XLABEL='',YLABEL='',SCALE=0,HEIGHT=5cm,WIDTH=5cm

ORIGIN=0cm,1.5cm
PLOT LY*LX/xmin=5,XMAX=9.95,ymin=5,YMAX=9.95,
   XTICK=5,YTICK=5,SYMBOL=2,SIZE=RELSIZ,LABEL=SUBTAG$,
   XLABEL='',YLABEL='',SCALE=0,HEIGHT=5cm,WIDTH=5cm

ORIGIN=5cm,1.5cm
PLOT LY*LX/xmin=10,XMAX=14.95,ymin=5,YMAX=9.95,
   XTICK=5,YTICK=5,SYMBOL=2,SIZE=RELSIZ, LABEL=SUBTAG$,
   XLABEL='',YLABEL='',SCALE=0,HEIGHT=5cm,WIDTH=5cm

ORIGIN=10cm,1.5cm
PLOT LY*LX/xmin=15,XMAX=20,ymin=5,YMAX=9.95,
   XTICK=5,YTICK=5,SYMBOL=2,SIZE=RELSIZ, LABEL=SUBTAG$,
   XLABEL='',YLABEL='',SCALE=0,HEIGHT=5cm,WIDTH=5cm

SELECT
SELECT LY>=10

ORIGIN=-5cm,6.5cm
PLOT LY*LX/xmin=0,XMAX=4.95,ymin=10,YMAX=14.95,
   XTICK=5,YTICK=5,SYMBOL=2,SIZE=RELSIZ, LABEL=SUBTAG$,
   XLABEL='',YLABEL='',SCALE=0,HEIGHT=5cm,WIDTH=5cm

ORIGIN=0cm,6.5cm
PLOT LY*LX/xmin=5,XMAX=9.95,ymin=10,YMAX=14.95,
   XTICK=5,YTICK=5,SYMBOL=2,SIZE=RELSIZ, LABEL=SUBTAG$,
   XLABEL='',YLABEL='',SCALE=0,HEIGHT=5cm,WIDTH=5cm

ORIGIN=5cm,6.5cm
PLOT LY*LX/xmin=10,XMAX=14.95,ymin=10,YMAX=14.95,
```

Program 3.2.4. (continued)

```
            XTICK=5,YTICK=5,SYMBOL=2,SIZE=RELSIZ, LABEL=SUBTAG$,
            XLABEL='',YLABEL='',SCALE=0,HEIGHT=5cm,WIDTH=5cm

        ORIGIN=10cm,6.5cm
        PLOT LY*LX/xmin=15,XMAX=20,ymin=10,YMAX=14.95,
            XTICK=5,YTICK=5,SYMBOL=2,SIZE=RELSIZ, LABEL=SUBTAG$,
            XLABEL='',YLABEL='',SCALE=0,HEIGHT=5cm,WIDTH=5cm

        ORIGIN=-5cm,11.5cm
        PLOT LY*LX/xmin=0,XMAX=4.95,ymin=15,YMAX=20,
            XTICK=5,YTICK=5,SYMBOL=2,SIZE=RELSIZ, LABEL=SUBTAG$,
            XLABEL='',YLABEL='',SCALE=0,HEIGHT=5cm,WIDTH=5cm

        ORIGIN=0cm,11.5cm
        PLOT LY*LX/xmin=5,XMAX=9.95,ymin=15,YMAX=20,
            XTICK=5,YTICK=5,SYMBOL=2,SIZE=RELSIZ, LABEL=SUBTAG$,
            XLABEL='',YLABEL='',SCALE=0,HEIGHT=5cm,WIDTH=5cm

        ORIGIN=5cm,11.5cm
        PLOT LY*LX/xmin=10,XMAX=14.95,ymin=15,YMAX=20,
            XTICK=5,YTICK=5,SYMBOL=2,SIZE=RELSIZ, LABEL=SUBTAG$,
            XLABEL='',YLABEL='',SCALE=0,HEIGHT=5cm,WIDTH=5cm

        ORIGIN=10cm,11.5cm
        PLOT LY*LX/xmin=15,XMAX=20,ymin=15,YMAX=20,
            XTICK=5,YTICK=5,SYMBOL=2,SIZE=RELSIZ, LABEL=SUBTAG$,
            XLABEL='',YLABEL='',SCALE=0,HEIGHT=5cm,WIDTH=5cm

        END
    ENDTEXT
    SKIP
ENDDO

SET TALK ON
SET SCOREBOARD ON
SET SAFETY ON
CLEAR ALL
CLEAR
```

Box 3.2.4. Program PRINTMAP.PRG for creating the field map for a recensus.

Program 3.2.4 is a FoxPro program which creates a series of Systat graphics programs for drawing maps for a recensus. FoxPro cannot create the complex graphs necessary, and Systat has a more complete and versatile graphics module (for instance, it will automatically place tag numbers next to stems marked on the map). It is written for an HP Laser Jet 4 printer, and although it will work on other printers, some of the fonts, characters, and positions in the Systat program may have to be adjusted.

There are two sections of Program 3.2.4. The first part creates a temporary FoxPro data set for each 20x20 m quadrat in the plot, including all the data necessary for making the maps. This section of the program is pure FoxPro, with no Systat involved. In the second part of the program, FoxPro code is used to write text to files, creating data files and programs to be used by Systat. One data file and one program are created for each quadrat.

PRINTMAP.PRG starts by filling a new database TEMPDATA.DBF. Before executing, TEMPDATA.DBF must be created and given all the same fields as the main census database. The fields are then copied from the main census database (line 8). In addition, four new fields must be built into TEMPDATA.DBF: *SUBTAG* is the last three digits of each tree's tag number, plus the code 'R' or 'D' for resprouts and dead trees respectively (created with lines 10-12). This is what will be printed on the map next to every plant. (Printing only the last three digits of the tag, rather than the entire number, reduces clutter on the map but still allows trees to be located.) *LX* and *LY* are local coordinates within a 20x20 m quadrat, and are calculated from *GX* and *GY* in the main database (line 13). This is the opposite of the steps taken in Program 3.1.4, APPMAP.PRG, to convert local coordinates to global coordinates (see Fig. 3.1.2). *RELSIZE* is the variable used to indicate how big the circle representing each tree on the map will be, and is created by lines 16-27. All trees < 100 mm dbh, dead trees, and resprouts with no dbh, are given *RELSIZE* = 1, whereas trees ≥ 100 mm dbh are given a size proportional to dbh (line 22).

Next, the PRINTMAP program splits this large database into many smaller databases, one for each quadrat (lines 30-39). Each is named with the letter Q followed by the quadrat number, for instance, Q3513 DBF.

Beginning at line 43, the FoxPro program prints a Systat program as text to a file. *SET TEXTMERGE* is the FoxPro command for writing text to a file with the name SYSFIL.CMD (line 43). This command enables the merging of text with contents of database files and memory variables. After *SET TEXTMERGE*, lines followed by \\ or \ are printed character for character (the latter is preceded by a carriage return). In addition, any lines between *TEXT* and *ENDTEXT* are printed character for character. Hence, lines 41-55 create a file with two lines for each quadrat (these are the first two sets, for quadrats 0000 and 0001):

```
SAVE Q0000
IMPORT "Q0000.DBF"/TYPE DB3
SAVE Q0001
IMPORT "Q0001.DBF"/TYPE DB3
```

etc. The command *INDEX ... UNIQUE* (line 42) forces FoxPro to skip through the database, picking out only the first record for each quadrat, so the program prints commands for each quadrat just once.

SAVE and *IMPORT* are the Systat program language commands for importing a FoxPro database (Q0000.DBF) and saving it as a Systat file (Q0000.SYS). The remainder of the program, starting at line 59, is FoxPro code which creates the large set of Systat programs which actually draw the maps. For each quadrat, the variable *mF* is assigned the quadrat number plus a Systat extension, *.CMD*; this will be used to name each Systat program (lines 62-64). Thus, the first text file created is called Q0000.CMD, the second Q0001.CMD, etc. Again, since the index *Q20* was created with *UNIQUE*, one CMD file is created for each quadrat, and one map is printed per quadrat. All lines starting with 65, and then 67-208, represent the text that is printed character-for-character into the new Systat command file.

Lines 67-119 are Systat code which print a series of header information on each map. *ORIGIN* indicates where to print, and *WRITE* simply prints what follows. The \3, *h*, and *w*, are Systat commands to set the font, reduce height of the font by 50%, and reduce width of the font by 50%. The *HEIGHT* and *WIDTH* commands set the default font size.

Lines 121-206 print the map itself, a series of boxes for each 5 x 5 m subquadrat and a small circle with the subtag number for each tree. *USE* names the database, and *CS* sets the character size for the graph (lines 121-122). The command *SELECT LY<10* is necessary when there are many plants in a 20x20 quadrat—it tells Systat to store just those records with *LY<10* into memory. This is not necessary for sites with lower stem density. Then follow a series of *ORIGIN* commands to place each subquadrat and *PLOT* commands to place a dot at *LX, LY* for each tree. Following *PLOT* are a series of commands that describe the graph; most important, *LABEL* prints the subtag next to each tree. (*PLOT* is repeated once per subquadrat simply because of memory limitations.) Lines 165-166 deselect the first set of trees, then select those with *LY≥10*, and the *ORIGIN* and *PLOT* commands are repeated for the eight upper subquadrats (lines 168-206). Then *END* completes the Systat program (line 208), and *ENDDO* causes FoxPro to return to line 61 and start work on the next quadrat.

After PRINTMAP.PRG is executed, there will be 1250 FoxPro databases (each holding data from one quadrat), a single Systat program called SYSFIL.CMD, and 1250 more Systat command programs called Q0000.CMD through Q4924.CMD (one for each quadrat). SYSFIL.CMD is executed next, and this opens each FoxPro database in succession and saves it as a Systat database (Q0000.SYS through Q4924.SYS). Then the remaining 1250 Systat programs must be executed; each opens one Systat database and prints the map for the corresponding quadrat. The 1250 programs are executed by running a DOS batch file from within Systat. The batch contains 1250 lines:

```
submit Q0000.CMD
submit Q0001.CMD
submit Q0002.CMD
..................

..................
submit Q4922.CMD
submit Q4923.CMD
submit Q4924.CMD.
```

Program 3.2.5. Program for creating field data sheets for recensus.

Program PRINTPG.PRG

```
SET TALK OFF
SET SCOREBOARD OFF
SET SAFETY OFF
CLEAR
CLEAR ALL

SELECT 1
USE TEMPDATA
INDEX ON Q20+P5+TAG TO Q20P5TAG

DO WHILE NOT EOF()

   mQ20=Q20
   mP5=P5
   gorec=RECNO()
   COUNT TO nostems WHILE Q20=mQ20 AND P5=mP5

   IF MOD(nostems,3)>0
      mROWS=INT(nostems/3)+1
   ELSE
      mROWS=INT(nostems/3)
   ENDIF

   GO gorec
   DO WHILE Q20=mQ20 and P5=mP5
      norow=1
      DO WHILE norow<=mROWS and Q20=mQ20 and P5=mP5
         REPL ROW WITH norow
         norow=norow+1
         SKIP
      ENDDO
   ENDDO

ENDDO
CLEAR ALL

SET DEVICE TO PRINTER
drline=CHR(95)+CHR(95)+CHR(95)+CHR(95)+CHR(95)+CHR(95)+CHR(95);
   +CHR(95)+CHR(95)+CHR(95)+CHR(95)+CHR(95)+CHR(95)+CHR(95);
   +CHR(95)+CHR(95)+CHR(95)+CHR(95)+CHR(95)+CHR(95)+CHR(95);
   +CHR(95)+CHR(95)+CHR(95)+CHR(95)+CHR(95)+CHR(95)+CHR(95);
   +CHR(95)
```

Program 3.2.5. (continued)

```
USE TEMPDATA
INDEX ON Q20+P5+STR(ROW,2)+TAG to Q20P5RTG

DO WHILE NOT EOF()
   pgno=1
   mQ20=Q20
   mcol=SUBSTR(Q20,1,2)

   DO WHILE SUBSTR(Q20,1,2)=mcol

      mQ20=Q20
      @1,1 SAY SUBSTR(mQ20,1,2)+','+SUBSTR(mQ20,3,2)
      @1,33 SAY "FOREST DYNAMICS PROJECT"
      @1,81 SAY SUBSTR(mQ20,1,2)+','+SUBSTR(mQ20,3,2)+'
         ('+ALLTRIM(STR(pgno,2))+')';
      @3,30 SAY "Please write your name and date:"
      @4,1 SAY "Names:"
      @4,31 SAY "1. Data entry by:"
      @4,61 SAY "2. Data entry by:"
      @5,1 SAY drline
      @5,31 SAY drline
      @5,61 SAY drline
      @6,1 SAY drline
      @6,31 SAY drline
      @6,61 SAY drline
      @7,1 SAY drline+drline+drline+CHR(95)+CHR(95)

      gorec=RECNO()
      COUNT TO mtotq20 WHILE Q20=mQ20
      GO gorec
      mL=pROW()+1

   DO WHILE Q20=mQ20

      mP5=P5
      gorec=RECNO()
      mRT=ROW
      SKIP
      DO WHILE Q20=mQ20 AND P5=mP5
         IF ROW>mRT
            mRT=ROW
         ENDIF
         SKIP
      ENDDO
      GO gorec

```

Program 3.2.5. (continued)

```
        IF mL+1+mRT>=64
            @mL,40 SAY “Continued”
            EJECT
            pgno=pgno+1
            @1,1 SAY SUBSTR(mq20,1,2)+','+SUBSTR(mq20,3,2)
            @1,33 SAY “FOREST DYNAMICS PROJECT”
            @1,81 SAY SUBSTR(mq20,1,2)+','+SUBSTR(mq20,3,2)+' (';
                +ALLTRIM(str(pgno,2))+')'
            @2,1 SAY drline+drline+drline+CHR(95)+CHR(95)
            mL=3

        ELSE
            @mL,1 SAY SUBSTR(mP5,1,1)+','+SUBSTR(mP5,2,1)
            mL=mL+1
            DO WHILE P5=mP5 and Q20=mQ20
                mR=ROW
                @mL,1 SAY TAG+' '+MNEM+' '+STR(DBH,4)+CODE
                @mL,23 SAY CHR(95)+CHR(95)+CHR(95)+' ';
                        +CHR(95)+CHR(95)+CHR(95)
                SKIP
                IF ROW=mR and P5=mP5 AND Q20=mQ20
                    @mL,31 SAY TAG+' '+MNEM+' '+STR(DBH,4)+CODE
                    @mL,53 SAY CHR(95)+CHR(95)+CHR(95)+' ';
                        +CHR(95)+CHR(95)+CHR(95)
                    SKIP
                ENDIF
                IF ROW=mR AND P5=mP5 AND Q20=mQ20
                    @mL,61 SAY TAG+' '+MNEM+' '+STR(DBH,4)+CODE
                    @mL,83 SAY CHR(95)+CHR(95)+CHR(95)+' ';
                        +CHR(95)+CHR(95)+CHR(95)
                    SKIP
                ENDIF
                mL=mL+1
            ENDDO
            @mL,1 SAY drline+drline+drline+CHR(95)+CHR(95)
        ENDIF

        mL=mL+1

    ENDDO

    @mL,20 SAY “END OF QUADRAT: There are “;
            +ALLTRIM(STR(mtotq20,6))+;
            “ stems in Quadrat “+SUBSTR(mQ20,1,2)+', ';
            +SUBSTR(mQ20,3,2)+'.'
    EJECT
```

Program 3.2.5. (continued)

```
138          pgno=pgno+1
139
140       ENDDO
141    ENDDO
142
143    CLEAR ALL
144    SET TALK ON
145    SET SCOREBOARD ON
146    SET SAFETY ON
147    SET DEVICE TO SCREEN
```

Box 3.2.5. Program PRINTPG.PRG for creating recensus field sheet.

Program 3.2.5 creates a field sheet with a list of all trees in a given quadrat, including prior dbhs and codes. It is written entirely in FoxPro. To use this program, a new, temporary database must be created which has all records from a prior census, with the fields *TAG, Q20, P5, DBH,* and *CODE.* In addition, a new blank field named *ROW* must be created. The new database is called TEMPDATA.DBF (line 8). The output of this program will be the data sheet shown in Figure 2.5.2. Notice that all plants from a single 5x5 m subquadrat are printed in a group, in three columns across the page. Tags from one subquadrat are printed in sequence, down the left column, then down the center column, then down the right column. Then the next subquadrat is printed below. This is the most useful printing pattern for the field workers.

The first section of the program (lines 11-35) determines how many rows are needed for printing records of each subquadrat. This is necessary because of the way rows and columns are printed. First, the number of records in a subquadrat are counted into the variable *nostems* (line 16). Lines 18-23 determine how many rows are needed for this subquadrat by dividing by three. Lines 24-31 determine the row number on which each record will be printed (within a subquadrat) and store this permanently in the new field, *ROW.* This requires the double *DO-WHILE* loop, which cycles through all records in the subquadrat (line 25), and within this, through all records in each successive column (line 27). When the row number *norow* reaches the maximum *mROWS*, the inner loop ends and the next column is begun. For example, if a subquadrat has 17 plants, then lines 18-22 will determine that six rows are needed. Lines 24-32 will assign the first six tags in the quadrat to rows 1-6, the next six tags to rows 1-6, and the final five to rows 1-5.

The rest of the program prints the page. A variable *drline* is created to store a line about one-third the width of the page (lines 40-44); CHR(95) returns the symbol for a single underscore character. Then the data set is re-indexed, this time on quadrat, subquadrat, row, and tag. A loop through the entire data set is begun (line 49), and the variable *pgno* is printed at the top of the form. The second loop which begins at line 54 runs through each column of quadrats, so that the page number is reset to 1 at the end of each column. The first section, lines 56-71, print header information at the top of a page where a given quadrat begins. This includes the quadrat number (lines 57 and 59) and a series of blank lines for field workers to enter names and dates (lines 61-71).

Lines 73-75 count the number of records in the current quadrat and return to the start of the quadrat. Line 76 determines the current line in the printer with the function *pROW()*; the variable *mL* will keep track of the line number. Then a loop through the current quadrat is begun to finish printing all its records (line 78).

Lines 84-90 loop through a single subquadrat and find the highest row number within, then return to the start of that subquadrat. Line 92 tests whether there are enough rows left on the page to print that entire subquadrat (so trees from a single subquadrat are never printed across two pages). If there is not enough space, lines 93-95 print a comment indicating the end of the page, eject the page, and augment the *pgno* variable. Then lines 96-101 print the header on the next page, including the next page number. Since this is the second page of the quadrat, blanks for names and dates are not needed.

If there is enough space, then lines 103-127 print data from all plants in that subquadrat. First, the subquadrat number is printed (line 104). Then the line counter advances one (line 105) for the data. The fields *TAG*, *MNEM*, *DBH*, and *CODE* are printed in the left-hand column (at position *mL,1*; meaning line *mL*, position 1), and two blank lines are printed to the right, where data will be recorded (at position *mL, 23*). Line 111 moves to the next record, then 112 tests whether this record is to be printed on the same *ROW* as the prior record. If so, it is printed at *mL*—the same line—but at position 31, which is one-third of the way across the page. If the subsequent record is still on the same *ROW*, then it is printed at position 61, in the third and last column. But when a record is found with the next *ROW*, *mL* is advanced by one (line 124), control is returned to the start of the loop at line 107, and printing begins in the leftmost column. Because the records in each subquadrat are indexed first on *ROW*, then on *TAG*, successive records are not successive tag numbers. For example, the first row printed might contain tags 1, 6 and 11, the next row tags 2, 7 and 12, etc. When a subquadrat is finished, a line is drawn to separate it from the next (line 126).

When the quadrat is finished, the loop ends at line 131, and the number of stems in the quadrat is printed at the end (lines 133-136). This finishes the quadrat loop (line 140), and printing of the next quadrat begins on a fresh page (line 78 again).

CHAPTER 3.3

Constructing a Database After the Second Census

Creating the database after a second or subsequent census involves many of the same steps that were taken for the first census database. However, there are a number of new wrinkles, and some of these become complex. Death and recruitment must be considered, and particularly error-prone is the issue of the point-of-measure (the POM) on multiple-stemmed plants, resprouts, and buttressed trees.

Data from a recensus is entered into computer files which are then merged and corrected, as described in chapter 3.1. Four different FoxPro databases are thus produced: one for recruits (those newly tagged in the recensus), one for original trees (tagged in an earlier census), plus those for multiple stems and big trees (Table 3.3.1). Three of these have precise analogs from the first census (Table 3.2.1) and are handled just like the first census databases. The original tree database is different, however.

Recruits

The file for recruits is exactly analogous to the main database created from the first census, and it can be handled exactly as described in chapter 3.2. Dates and names of people are shifted to a quadrat database and then removed from the recruit database. Dbhs for multiple-stemmed plants are taken from the multiple-stem database, and likewise for big trees. Stem number must be counted from the multiple-stem database, and new fields for POM and status code are added. Just as for the first census, any new plant in a second census is automatically given *POM* = 1 and *status code* = 'A'. FoxPro programs for each of these steps were given in chapter 3.2. Finally, the database is divided into two—tag, species name, quadrat, subquadrat, and coordinates into a permanent database, and tag, dbh, POM, status code, stem number, date, and general codes into a census database. These will later be appended to the databases for original trees to create the complete database.

Original Trees

When the data collected on original trees in a recensus are entered into the computer (chapter 3.1.1), they end up in a database that is similar to that for recruits. Thus, several of the steps taken with the original-tree database are identical to those taken with the recruit database. Dates and names for each quadrat are shifted to a quadrat database and then removed from the original-tree database.

Tropical Forest Census Plots: Methods and Results from Barro Colorado Island, Panama and a Comparison with Other Plots, by Richard Condit.

Table 3.3.1. Data fields in the four provisional databases from a large plot recensus. These were created by the programs given in chapter 3.1 on data entry.

A) Recruit database

data field	**contents**
person collecting data (initials)	2 characters
date censused	date
person checking data (initials)	2 characters
date checked	date
1st person entering data (initials)	2 characters
1st date entered	date
2nd person entering data (initials)	2 characters
2nd date entered	date
quadrat number	4-digit integer
subquadrat number	2-digit integer
tag number	6-digit integer
species name	four-letter code
x coordinate (m)	floating point number, 1 digit after decimal
y coordinate (m)	floating point number, 1 digit after decimal
dbh (mm)	4-digit integer
codes	up to 6 characters in a string

B) Original tree database

data field	**contents**
person collecting data (initials)	2 characters
date censused	date
1st person entering data (initials)	2 characters
1st date entered	date
2nd person entering data (initials)	2 characters
2nd date entered	date
quadrat number	4-digit integer
subquadrat number	2-digit integer
tag number	6-digit integer
species name	four-letter code
old codes	up to 6 characters in a string
dbh (mm)	4-digit integer
codes	up to 6 characters in a string

Table 3.3.1. (continued)

C) Multiple-stem database data field	contents
tag number	6-digit integer
dbh (mm)	4-digit integer

D) Big-tree database data field	contents
tag number	6-digit integer
dbh (mm)	4-digit integer
height of POM (m)	floating point number, 2 digits after decimal

Original trees with multiple stems must get dbhs from the multiple-stem database. Differences arise in handling big trees, dead trees and resprouts, however, and these govern the *status* and the *POM* data fields.

Dbh of Buttressed Trees

Initially, the dbh of big trees can be consulted from the big-tree database, as described in chapter 3.1. After this is finished though, it is necessary to go through and check for big trees that had more than two dbhs recorded. This happened when a large tree's buttresses moved upward and caused swelling at the first POM. In this case, field workers had to use a higher POM, but the dbh at the first POM was recorded as well, so there are two different dbhs in the big-tree database. Only one can be entered in the final database, and the database manager should decide this on a case-by-case basis (there should be very few trees in this category). If the dbh at the higher POM is entered in the main database, then the code 'C' must be added to the general codes (this is necessary for assigning the *POM* value, below).

The Status Code for Dead Trees

The status code is a single character meant to quickly and simply inform a user about whether a plant is alive or dead. In the first census, all plants were alive and thus received a status code 'A' (chapter 3.2), as were all recruits in a recensus. But the status code becomes useful when dealing with trees censused for the second time, which can get three different status codes. One of these codes is 'D', for dead. All dead plants get a status code 'D' in the final database for the recensus. They were given the general code 'D' by field workers. The 'D' is left in the general code field, as it does no harm there.

The Status Code for Resprouts

Resprouts are those plants which lost the stem measured in the first census but resprouted a new one. These were given the code 'R' by the field workers (chapter 2.5). As long as the resprouted stem is ≥ 10 mm dbh, these plants are alive, part of the census, and are given status code 'A'.

However, there is a subset of resprouts whose newly sprouted stem is < 10 mm dbh. They were given a code 'R' and a dbh 0 by the field workers, and they retain the general code 'R', but they have a unique status—they are alive and tagged and must be followed in the future, but they are technically not in the census because they are not ≥ 10 mm dbh. Indeed, if such a plant were encountered during a first census, it would be ignored. They are thus given a unique status code, 'F', meaning failure to reach the census.

Filling in the Status Code

To recap, most plants in the original database—all surviving plants with a stem ≥ 10 mm dbh—are given the status code 'A' in the final database. Dead trees get 'D' and living trees with no dbh ≥ 10 mm get 'F'.

Creating and filling a field with the status code is routine in FoxPro. It first requires that a new field be added by modifying the database structure, adding a variable called *STATUS*. Then the command *REPLACE ALL STATUS WITH 'D' FOR 'D'$CODE* fills that field with the new status code for dead trees. This is repeated for status 'A' and 'F' with the appropriate *FOR* statement (e.g., *REPLACE ALL STATUS WITH 'F" FOR 'R'$CODE .AND. DBH=0*).

Point-of-Measure (POM)

The new variable *POM* must also be added to the original-tree database. During the first census, and for recruits in a recensus, *POM* is always 1, but the value can vary for original trees. If a field worker decided that the old POM was invalid, and used a new one, he or she used the code 'C'. All original plants with a code 'C' have their *POM* incremented by 1 relative to the prior census; all remaining plants keep their original *POM*. Making this change thus requires consulting the earlier census database for original trees, which can be done by linking through the *TAG* variable.

Thus, each plant gets *POM*=1 in the census in which it was first recorded in the plot. Its *POM* remains 1 in all future censuses as long as its stem can be measured at the same spot. However, when a field worker finds that a new *POM* is needed, either because of a broken stem or a buttress or swelling, the *POM* is changed to 2. It remains at 2 for future censuses until another change is necessary. The purpose of the *POM* variable is to allow users to know whether growth can be calculated between any two censuses—growth would not be valid between any pair of censuses if the *POM* values differed.

At BCI, the code 'C' was not used through the 1995 census, and assigning POMs has been problematic. We adopted a series of methods which no longer make sense, but which I describe here fully in an Appendix to this chapter.

The Final Database

Splitting the Original Tree Database

Once the status code and POM have been added to the database of original trees, it can be divided into two—one database with permanent data and one with census data, exactly as was done with the recruit database. The permanent database for original trees is identical to the permanent database from the prior census, except in cases where corrections were made, in which case the newer version replaces the previous.

Merging Permanent Databases

There is now a permanent database for original trees and a permanent database for recruits. They are completely non-overlapping—they have no trees in common. They are merged into one large file that has tag number, species mnemonic, quadrat number, subquadrat number, and gx and gy coordinates, for all plants in either census. This creates the final permanent database.

Merging Recensus Databases

From the second census, there is one census database for original trees and one for recruits. As for the permanent databases, these are completely non-overlapping. They are appended into one large file that includes the tag, dbh, POM, stem number, status code, and general codes from the recensus for every plant. This creates the final database for the recensus.

Recreating the First Census Database

The census database from the first census has remained unchanged up to this point. It includes the tag, dbh, POM, stem number, status code, date, and general codes from the first census for all original trees. In most cases this information should not be changed; indeed, it should not be since it was collected five years earlier. There are some exceptions though. For example, if a tag number had to be changed on an original tree, it is necessary to go back and change the tag in the first census database.

This first census database should now be expanded with a step which at first glance seems rather peculiar, but which makes analyses much simpler and more efficient. All recruits from the second census should be appended to the first census database, with all columns but three left blank: the tag, the status code, and the date. The tag number must be inserted as the identifier. The status code is set to a new and different value, 'P', meaning prior, because these plants did not yet exist during the first census. This is the final status code, and every plant now has status 'D' for dead, 'A' for alive, 'F' for living with no dbh ≥ 10 mm, or 'P'. Finally, the date is filled in using the date on which each plant's quadrat was censused during the *first* census (not the recensus!); these dates are taken from the quadrat database from the first census. Giving a date may seem peculiar at first, but it actually makes perfect sense: on that date, the quadrat was censused and the plant was *not* found. The date is necessary for analyses of recruitment rate, which require time intervals.

Adding the recruits to the first census database involves three steps. First, the recruit database from the recensus is duplicated. The tag remains unchanged in the duplicated version, since it is the identifier for each record. 'P' is inserted into the status code for every record, and the date taken from the first census quadrat database. The remaining fields in the duplicated database are given blanks (a –1 for numerical fields and '.' for the general codes). Finally, this new database is appended to the first census database.

Complete Database

There are three databases now, one for permanent data and one for each of the two censuses. All three have exactly the same tag numbers—one for every single tree ever recorded in either census. The database from the first census includes mostly records of those trees mapped during the first census, with their first-census dbhs and full observations, plus all trees found in the second census with blank dbhs and status 'P'. The database from the second census includes all the same trees, most of which have dbhs filled in and status 'A', but dead trees have status 'D' and no dbh, while living trees with dbh < 10 have status 'F' and dbh 0. The permanent database simply has a list of every tag number used, with species name, quadrat, subquadrat, and coordinates.

Third and Later Censuses

Creating databases for subsequent censuses involves all the steps just described, but with a few minor considerations. For example, it is only in a third census that trees with status 'F'—alive but no stem ≥ 10 mm dbh—can recover their status 'A'. Thus, any plant that was status 'F' in the second census and grows to be larger than 10 mm dbh switches back to status 'A'. In addition, trees recorded as dead in one census sometimes show up alive again in the next. In these cases, status 'A' is assigned in the current census and the earlier census should be back-corrected to status 'A' (this is another example where a prior census database must be corrected).

New plants from the third census and beyond are always appended to each of the earlier census databases and given the status 'P'. Dead plants are also placed in every census, even long after they first died: even plants which died during the 1985 census at BCI are still listed with status 'D' in the 1995 census database. In both cases, these plants are assigned a census date as the day on which their quadrat was censused.

Every census database and the permanent database always include records for every tree ever censused—they all continue to expand as new censuses are finished.

To see why all "prior" plants and all dead plants should be listed in every census, consider an analysis one day well in the future, after the census of the year 2030 at BCI. Estimating mortality rate over a 48-year interval (1982-2030) will require tallying every plant that was alive in 1982 but dead in 2030, whether it first died in 1985, 1990, or 2025. A simple way to do this is to keep every plant that has ever died in the 2030 database; one then simply reads the 1982 database for all plants with status 'A', and the 2030 database to find all with code 'D'. If records of every dead plant were not maintained in each census database, it would be necessary to read every single census to find when each plant first died. This is also the reason that census date should appear for each tree—living, dead, or prior—in each census. The same considerations apply in calculating recruitment rates and growth rates; the *POM* variable makes growth calculations across long intervals equally easy.

Final Database Structure

The final structure of the multiple-census database is identical to that from the first census, summarized in Table 3.2.2, except there is now one census database for each census. For example, at BCI after 1995, we have five databases for the four censuses—the permanent database plus one for each census.

From these, we also created separate ASCII text files for each species. There are 315 species in the BCI plot, so 315 permanent files were created and stored in bci/census, plus 315 files from each census were created and stored in bci/census0 through bci/census3. There are thus 1575 text files for the four censuses. The text-file organization is diagrammed in Figure 3.3.1.

The multiple-stem and big-tree databases are maintained separate for each census. These files must be saved, since they record dbhs of minor stems or extra dbhs on big trees—those dbhs are not transferred to the main database. Multiple-stemmed trees from one census need not be the same as those from another—indeed, many will be different, and likewise for big trees.

As an alternative, instead of separating data from each census into a different database, a single FoxPro file that included data from all censuses together could be maintained. It would hold the tag, species, quadrat, subquadrat, coordinates, plus dbh, status, codes, POM, stems, and date for each census. This makes programming easier, however, the disadvantage is that it is much slower to run analyses if only one census is needed. Also, the file becomes ungainly, since each record would require 35 fields of data for five censuses. Text files could also be collapsed in this way to one large file per species.

Database Correction

Corrections in the Original Database

Certain errors caught during a recensus were set aside until now. These are errors detected in original plants which must be corrected not only in the current census, but in the prior census database as well: mistakes in the permanent data of a tree from the prior census. It is reasonable and not at all uncommon for field workers to detect mistakes in species identification, map coordinates, or tag, which were missed during the first census. They cannot be handled during normal data entry, because only the dbh and codes are entered for original trees. Thus, field supervisors compile a separate list of these errors, and they should be corrected now, after the full database is assembled, by the database supervisor. If a tag number has to be changed, both the permanent and all census files must be corrected. A record of all these changes must be saved for future reference.

Routine Screening

Most routine screening and error correction after a recensus can be done with procedures identical to those described for a first census (chapter 3.2). There is one new procedure that can be run though: growth rates can be examined for extreme values (chapter 2.5). This does not require special analytic programs. Once the census databases are completed, they are linked on tag, and growth for each tree estimated as the difference in dbh between censuses. This is stored to a new field, then growth in several size classes is ranked from highest to lowest, and the most extreme records are stored in a new database, printed, then checked in the field. These trees are re-measured, reasoning that many of the extreme records will be due to gross dbh errors (chapter 2.5).

Field Sheets for the Next Census

The fields sheets needed for another recensus are created from the database from the previous census, and this can be done just as described in chapter 3.2. There are a couple additional considerations as well. First, dead trees ought to be included on the main datasheets in the first census after they were found dead. That is, a plant recorded dead in the second census should be included on field

Table 3.3.2. Final multiple census database, ready for analysis. These are identical to those created after the first census (Table 2.1.2), except there is now more than one census database.

data field	contents
A) Permanent database	
data field	**contents**
tag number	6-digit integer
species name	four-letter code
quadrat number	four characters
subquadrat number	two characters
x coordinate (m)	floating point number, 1 digit after decimal
y coordinate (m)	floating point number, 1 digit after decimal
B) Census database (one for each census)	
data field	**contents**
tag number	6-digit integer
dbh (mm)	4-digit integer
POM number	2-digit integer
stem number	3-digit integer
status code	1 character
census date	date
general codes	up to 6 characters in a string
C) Multiple-stem database	
data field	**contents**
tag number	6-digit integer
dbh (mm)	4-digit integer
D) Big-tree database	
data field	**contents**
tag number	6-digit integer
dbh (mm)	4-digit integer
height of POM (m)	floating point number, 2 digits after decimal
E) Quadrat database	
data field	**contents**
quadrat number	4-digit integer
person collecting data (initials)	2 characters
date censused	date
person checking data (initials)	2 characters
date checked	date
1st person entering data (initials)	2 characters
1st date entered	date
2nd person entering data (initials)	2 characters
2nd date entered	date

```
/u/datasets/bci/ census/(all 315 species files)
                 census0/(all 315 species files)
                 census1/(all 315 species files)
                 census2/(all 315 species files)
                 census3/(all 315 species files)
```

Fig. 3.3.1. Diagram of the path structure for storing census data from the BCI 50 ha plot. This is drawn for a Unix computer, where the directory */u/datasets/bci* is for BCI data. Within that, there are five subdirectories; *census* holds the permanent data (tag number, quadrat, and subquadrat, and coordinates) for every individual, in 315 files (one for each species); *census0* holds data from the 1982 census (tag number, dbh, POM, status, stem number, codes, and date), again with one file for each species. The remaining three subdirectories hold parallel data for the 1985, 1990, and 1995 censuses, each with 315 files.

sheets for the third census (see chapter 2.5). It is useful to include the dead trees because occasionally they turn out to be alive. Dead trees need not be included in the second census after they die or beyond.

It is also very important that trees with status 'F' in the second census—those with no stem ≥ 10 mm dbh—be included on the maps and datasheets for the third census. These are the trees that lost their stem ≥ 10 mm dbh, but remained alive, and they are still part of the census. They must therefore be re-measured.

Key issues for multiple-census database

- The variable *POM* must be used to indicate when the point-of-measure changed, as shown by the code 'C'
- All recruits from the second census are added as "prior" records to the previous census
- Dead trees are stored in the database for every future census after their death
- Data are stored in one FoxPro file per census plus a file for permanent information
- Each database includes a record for every tag across all censuses
- ASCII text files are created as well, one per species per census plus one per species for permanent information
- Steps in creating the multiple-census database are:
  1) Assemble the recruit database into permanent and census databases, exactly as for main data in the first census
  2) Assemble the original tree database into permanent and census databases
  3) Add proper status code and POM to each census database
  4) Append recruit databases to original tree databases
  5) Append recruits to first census database

APPENDIX 3.3.

POM Problems at BCI

At BCI, changes in POM have caused considerable problems because the code 'C' was not used by field workers (through 1995), and paint marks have only been used on big trees. Difficulties arose in three circumstances that can force the POM to be changed: high buttresses, resprouts, and multiple-stemmed plants.

Big trees were measured around buttresses in 1982 (chapter 2.2). Since then, the POM was placed above but often very close to the buttresses, so that buttresses sometimes rose above the POM by the next census. However, the height of the POM was always recorded, so correctly assigning the POM in the database has been straightforward. Some large trees had dramatic shifts in diameter caused by a change in POM (especially 1982-85), but by use of the POM field in the database, we know to eliminate these from growth records.

In the case of resprouts and multiple stems, we have always been forced to infer changes from the presence of 'R' and 'M' codes. These are the rules we devised for assigning POM at BCI:

1) Any plant without 'R' (resprout), 'M' (multiple-stem), or 'B' (big-tree) got the same POM in the recensus as it did in the prior census. That is, without any information to the contrary, we assumed that the stem was normal and was measured at the same spot in both instances.
2) Any code 'R' assigned by a field-worker automatically meant a change of POM during the first census in which the 'R' appeared. If the 'R' appeared in consecutive censuses, the POM was considered to change the first time but not the second time unless the dbh declined by 25% or more the second time. Thus, a plant which was given 'R' for the first time in 1990 was switched from POM 1 in 1985 to POM 2 in 1990. If another 'R' appeared in 1995, the plant kept POM 2 unless the dbh declined substantially. The reason for this is that the code 'R' has generally been applied even if the break occurred before the previous census (chapter 2.5).
3) Multiple-stemmed plants were given the same POM in successive censuses unless the largest dbh declined by 25% or more. There were 29,056 multiple-stem plants in the 1990 census, and 1,925 had dbhs that shrunk more than 25%.
4) The POM was set to 0 in cases where no dbh was taken on a living tree, either if the tree was simply missed, or if a tree appeared dead but later proved not to be.

Future censuses at BCI will eliminate all these problems by marking the largest multiple stem, but mostly by empowering the field workers to make a decision, in the field, about whether a POM has changed. Thus this ad hoc set of rules need not be followed elsewhere.

CHAPTER 3.4

Data Availability

In this final chapter on the large plot database, I cover something completely different. Who has proprietary rights over the data, and who is allowed to use it? At first glance, this subject may appear to have nothing to do with the accuracy and usability of the data set, but in fact, it does. In some cases where intellectual property rights over the data are ill-defined, no one feels a personal commitment to the project. This can lead to a situation where the quality of the data suffers, because no one has a stake in it. If it is not clear who is responsible for correcting and updating the database, then multiple versions of the database may result, each differing from the others.

So on the one hand, a key goal of large, permanent forest plots is to make a large set of data available to other scientists, managers, or policy-makers with an interest in forest dynamics. From this perspective, it is important to make large-plot data readily available to others. On the other hand, the host institute or institutes and the chief scientists involved must maintain intellectual rights over the data in order to feel that they have a stake in the output. Balancing the twin needs of propriety and availability has been our goal with data from the CTFS network of large forest plots.

BCI Data

Copies of the BCI database are automatically distributed to the scientists who are in charge of the project every time a new edition is completed. The principal-investigators may change through time, of course.

Other scientists who wish to make use of the data set are given access to the complete database upon agreeing to the following terms. Researchers must state exactly how they will use the data, and what plans they have for publication, dissertation, or presentation; they may be denied access if someone else is already working on the same topic. In most cases, a database manager for CTFS at STRI provides the researcher with a particular subset of the data—one or a few individual species, for instances. He or she often provides analyzed data as well, such as growth or mortality rates. In some cases, we have made the entire database available to researchers by providing internet access to the computers which hold it.

Our policy at BCI continues to be that the data are the intellectual property of the principal investigators (PIs) and that publications which make use of the BCI database include all three as co-authors. There are exceptions though, and this depends on the extent to which the BCI data contribute to the article in question. The data can always be used in unpublished dissertations without listing the PIs as

Tropical Forest Census Plots: Methods and Results from Barro Colorado Island, Panama and a Comparison with Other Plots, by Richard Condit.

co-authors. In addition, in fields outside ecology or forestry science, researchers are sometimes permitted to publish articles using the data without including the BCI PIs as co-authors. This situation has arisen especially in economic or social sciences that can make use of information on forest productivity. The one overriding rule for use of the BCI data is that the PIs for the plot have final authority to grant use of the data and to determine authorship of publications.

Therefore, defining the principal investigators is obviously crucial. This clearly must be agreed upon in advance by participating scientists. With the BCI project, the PIs are those who develop the research protocol, raise the money, and hire, train, and supervise assistants. A sticky question is how this changes through time. The group of people defined as principal investigators can shift: we have set an arbitrary and loose rule that when a scientist ceases participation with the project for several years, then he or she is no longer a PI, and likewise, when a new scientist joins at the PI level, he or she is.

To ensure that scientists who seek access to the BCI data understand our approach, a standardized form is used. A sample is shown in Figure 3.4.1. It describes the rules outlined above, and requires a description of how the data will be used. The interested scientist must agree to the conditions by signing the form. One of the principal investigators on the project should be responsible for data requests, and the others kept informed.

Between 1990 and 1997, 32 researchers were granted access to part or all of the BCI plot data. Of those, most wanted to do field work in the plot, but four wished only to work on the data set from afar. The only requests that were turned down were cases where there was no specific use for the data given with the request.

The Other Large Plots

For the most part, the principal investigators for the other large plots that are part of the CTFS network all maintain the same intellectual property rights to their own data sets as we do at BCI. Data can only be used after getting permission from the PIs. PIs from each plot have agreed to leave copies of their data sets at STRI in Panama to facilitate comparative studies, but use of a data set in Panama still requires permission from the PIs of the plot in question. Thus far, requests to use data from other plots have almost always been granted, and several comparative studies have been carried out (Condit et al 1996c, 1998; Hall et al 1998).

Request for Use of Data from the BCI 50 Ha Plot

I request use of all or part of the database from the 50 ha forest dynamics plot at Barro Colorado Island. I understand that in order to get access to the data, I must explain the specific use I intend for it as well as a list of publications or other forms of presentations I anticipate resulting from my work. I also understand that for publications in scientific journals which rely heavily on the BCI data, the principal investigators may be included as co-authors, at their request.

The principal investigators of the forest dynamics plot at BCI reserve the right to deny access to the data, and to decide who should be included as co-authors on any paper.

On the back of this sheet, I provide a one paragraph description of my intended use of the data, including publications that I anticipate (this is not a commitment to writing the papers, though). I understand that I must contact the database manager for the BCI plot in order to get the data I need, along with its documentation, once my request is granted.

Signature ______________________________ Date __________

Fig. 3.4.1. Sample form for requesting access to data from the large forest plot at Barro Colorado Island, Panama. This generalized form does not give names or addresses of the principal investigators nor the database manager, but the actual form currently in use does give specific names. These names can change over the years, however, hence the general nature of this form.

Part 4: Species Distribution Maps from the 50 ha Plot at Barro Colorado Island

Part 4

Introduction

Here I present results of the big census at Barro Colorado: maps showing the distribution of individual tree species in the 50 ha plot. I selected 18 maps to illustrate typical distributional patterns. Included are common species, rare species, widespread species, species occurring only in specific habitats (habitat specialists), species occurring in all habitats (habitat generalists), and species with very uneven (or patchy) distributions. For each species, all individuals ≥ 10 mm dbh are shown. To help understand the tree distributions, I also provide a habitat map for the 50 ha plot.

Habitat Map

Figure 4.1 is a map of the 50 ha plot, 1000 m long (east-west) by 500 m wide (north-south). The small squares on the map are 20 x 20 m quadrats. Topographic and edaphic habitats are indicated by various shades of black, gray, and white, and there are contour lines at 2 m vertical intervals; these same lines appear on each species map. Condit et al (1995a, 1996b) and Harms (1997) provide descriptions and definitions of the habitats.

Near the north boundary of the plot, to the right of center, is a section of young forest, shaded with black in Figure 4.1. This area was cleared of forest by French settlers around the turn of the century. The young forest and the area south of it, between 100 and 500 m north-south and 600 and 800 m east-west is the high plateau, which is the top of Barro Colorado Island; it is shaded with the darkest gray on the map. West of this is the largest habitat block in the plot, the low plateau, shaded with the next lighter shade of gray. It occupies most of the western half of the plot as well as a narrow strip east of 900 m. There are also a few isolated quadrats of low plateau near the south boundary. The high and low plateaus are defined as all quadrats with an inclination < 7° and with no standing water nor streams. The plateau is the driest part of the plot because it is furthest from the water table, and the high plateau is especially dry.

The slope habitat is where the summit of Barro Colorado Island slopes away from the plateau. It is defined as all quadrats inclined at ≥ 7°, and it is shaded with the lightest gray in Figure 4.1. There are slopes between 800 and 900 east-west, along the south boundary of the plot, and in the northwest and southwest corners. The water table below the plot reaches the surface at these slopes, so they are relatively wet compared to the plateau (Becker et al 1988; Condit et al 1995a; Condit et al 1996b). Indeed, there are several permanent springs on the slopes which seep water throughout the dry season, and one feeds a stream in the far northeast corner of the plot. The black quadrats outlining the high plateau are those having a mixture of high plateau and slope habitat.

Tropical Forest Census Plots: Methods and Results from Barro Colorado Island, Panama and a Comparison with Other Plots, by Richard Condit.

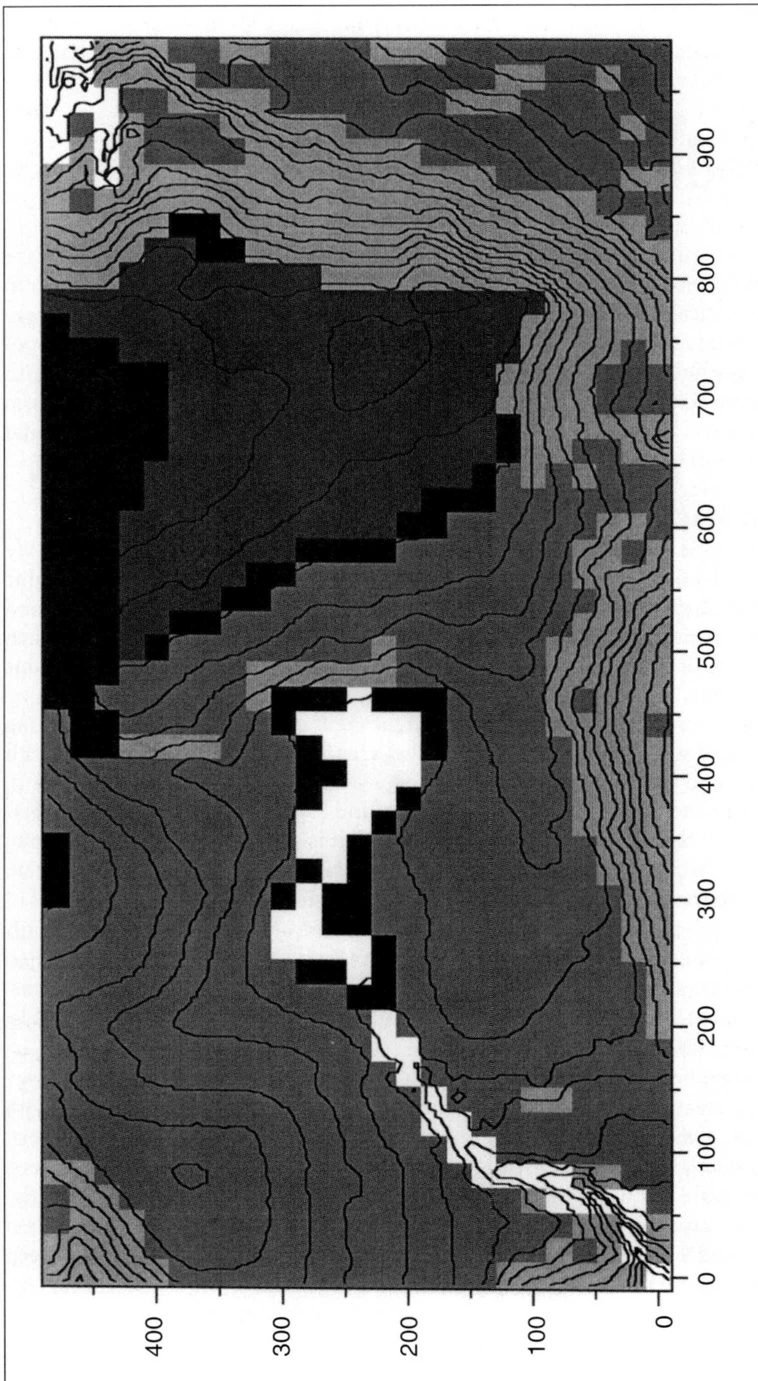

Fig. 4.1.

The region indicated in white on the map, near the center of the plot (250 to 450 m east-west, 200 to 300 m north-south), is a seasonal swamp. It usually has standing water throughout the wet season and remains wet in the dry season. On the map, the swamp is surrounded by a ring of black quadrats, those having a mixture of swamp and low plateau habitat. The other white quadrats on the map are adjacent to streams that flow during the wet season. One drains southwest out of the swamp, and there are two smaller ones that merge and then flow out of the northeast corner of the plot.

Species Maps

Figure 4.2 consists of 18 species distribution maps, each with the same contour lines as shown in Figure 4.1. All plants alive in 1982 are mapped. Those still alive in 1995 are indicated by crosses and solid symbols, according to dbh class (as indicated on the legend). Those that had died by 1995 are shown with open circles. There are six sizes of open circles, corresponding to plants from the six dbh classes shown on the legend. Each size open circle matches the size of the symbol for living plants in the same dbh class.

Generalist Species

Habitat generalists are defined as those whose distributions show no association with the habitats of the plot (Fig. 4.1). The most striking generalists are those that occur throughout the 50 hectares at a fairly uniform density, and there are many species showing this pattern. The first three maps (Fig. 4.2A to 4.2C) are examples: *Brosimum alicastrum* is a common tree, *Ocotea cernua* is uncommon, and *Ceiba pentandra* is sparse.

Ceiba illustrates another interesting pattern. Although there are very few individuals, most are immense trees, and this species actually includes a dominant share of the forest biomass. There are seven species whose populations include a few large trees and virtually no juveniles in the 50 ha plot.

Patchy Generalists

Many species in the plot have strongly patchy distributions, that is, show wide variation in abundance. Many such distributions are not associated with habitat features, so they also qualify as habitat generalists.

The most striking of these patchy species is an understory shrub, *Anaxagorea panamensis* (Fig. 4.2D). There are over 600 individuals of this species—every one in just a single hectare. It completely dominates the understory of that hectare. This species is only known from 4 or 5 patches like this in the entire world, all within a few kilometers of Barro Colorado Island.

Quite a few species show patchiness somewhat less distinctly than *Anaxagorea*. Two examples are *Xylopia macrantha* and *Coussarea curvigemmia* (Figs. 4.2E,F). Rare species such as *Hirtella americana* are frequently patchy too, occurring in small clumps (Fig. 4.2G). There is no indication that these species associate with one another, further evidence that the distributions are not explained by habitat features.

Habitat Specialists

These are species clearly associated with the habitat features of the plot. As such, they are patchy, since they occur in some places but not others. Unlike the patchy generalists though, distributions of these species are concordant, illustrating that they follow the same habitat features.

Most interesting are the slope specialists, because these distributions were not anticipated when the plot was first mapped. No one knew then that the slopes were moist, as the difference is not obvious from just walking in the forest. There are about 30 such slope specialists, and *Chrysoclamys eclipes*, *Guatteria dumetorum*, and *Poulsenia armata* are clear examples (Figs. 4.2H,I,J). Notice how similar the three distributions are—very abundant on the slopes in the east and south, sparse on the high plateau and around the swamp, and moderately abundant on the low plateau west of the swamp (review again Fig. 4.1 showing the habitats). Nearly all the slope-specialists have suffered high mortality and population decline since 1982, most likely because of the severe droughts that Barro Colorado has suffered over the past three decades (Condit et al 1996a). *Conostegia cinnamomea* provides a striking example of high mortality (Fig. 4.2K).

The American oil palm, *Elaeis oleifera*, is a clear swamp-specialist, completely confined to the area of standing water (Fig. 4.2L). There are a few other species which are most abundant in the swamp but occur elsewhere in smaller numbers, such as *Psychotria grandis* (Fig. 4.2M).

Faramea occidentalis is an unusual habitat specialist. It occurs most commonly on the driest parts of the plot, clearly avoiding the swamp and less clearly avoiding the slopes (Fig. 4.2N). Notice that although there are lots of *Faramea* on the slopes, the density is much greater on the high plateau.

Many of the abundant species in the plot, such as *Alseis blackiana*, avoid the swamp (Fig. 4.2O). *Alseis* is also one of the few species that is strikingly more abundant in the patch of secondary forest in the north-central part of the plot. Its distribution almost exactly demarcates this patch (see Fig. 4.1).

There are a few moisture-demanding species which occur only on the banks of the small streams and seeps. *Marila laxiflora* is an obvious example (Fig. 4.2P). The small group of individuals on the south boundary are at a little spring, and the remaining are along the larger streams at the northeast and southwest corners (see Fig. 4.1).

Gap Specialists

One final kind of habitat specialist are light-demanding pioneers, such as *Croton billbergianus* (Fig. 4.2Q). The most extreme light-demanders typically occur in small clumps in canopy gaps, where large trees have fallen and light reaches to the lowest levels of the forest. These gaps are ephemeral, appearing and disappearing as canopy trees die and then branches and trees grow to fill in the opening. As a result, entire clusters of pioneers grow up and then die together, as can be clearly seen in *Croton*'s distribution. Most pioneers suffer high mortality (Condit et al 1995a), but they also recruit well; *Croton* is actually increasing in abundance despite its high mortality.

Rare Species

There are many rare species in the Barro Colorado plot. In 1995, 23 species were represented by a single individual, and 44 more had 2-9 individuals. Figure 4.2R illustrates the "distribution" of *Ormosia amazonica*, one rare species. There was a single tree in the plot in 1982.

The Full Dataset

I have similar maps printed for all 315 species in the BCI plot, and interested readers can contact me for more. The complete species list can be found in Condit et al (1996b) or on the Center for Tropical Forest Science web site (http://www.si.edu/organiza/centers/stri/forest/ctfs.htm). The original dataset giving coordinates, species name, and diameter for all 326,000 individuals that have been tagged will also be made available to anyone interested in delving more deeply into the data.

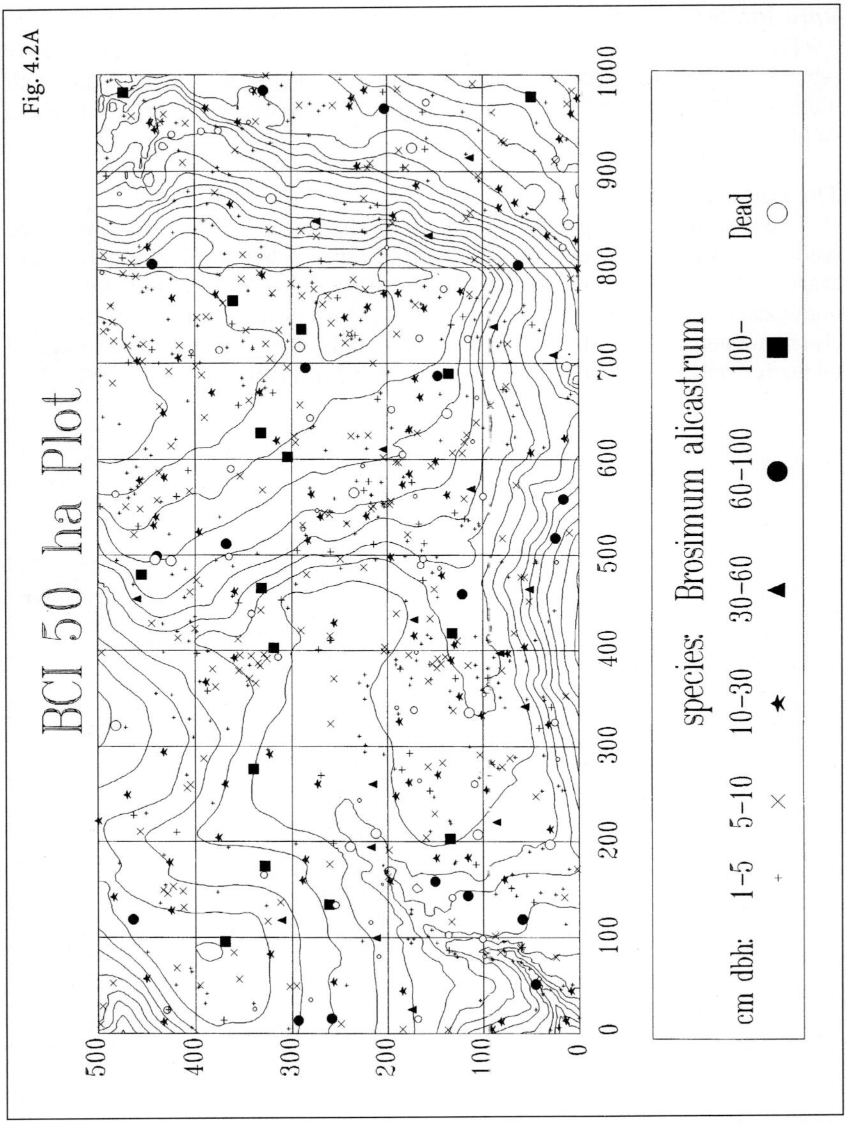
Fig. 4.2A
BCI 50 ha Plot
500
400
300
200
100
0
0
100
200
300
400
500
600
700
800
900
1000
species: Brosimum alicastrum
cm dbh: 1–5 5–10 10–30 30–60 60–100 100– Dead

Fig. 4.2B

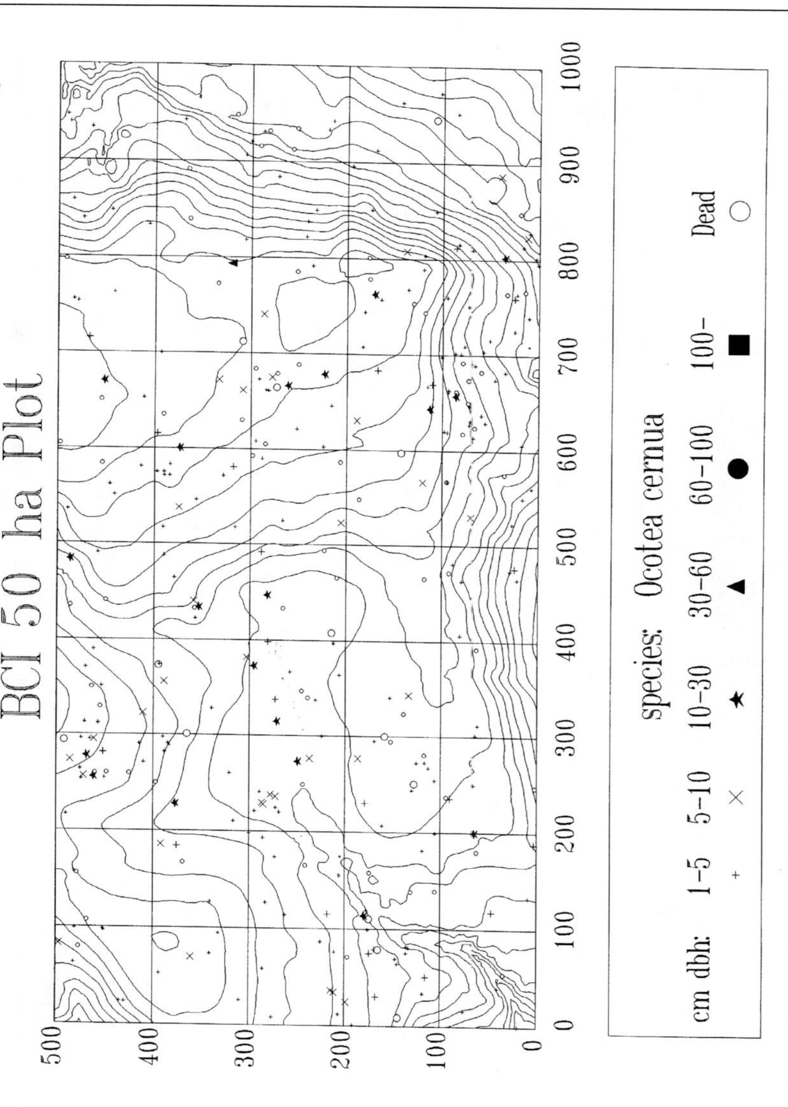
BCI 50 ha Plot
0
100
200
300
400
500
600
700
800
900
1000
species: Ocotea cernua
cm dbh: 1–5 5–10 10–30 30–60 60–100 100– Dead

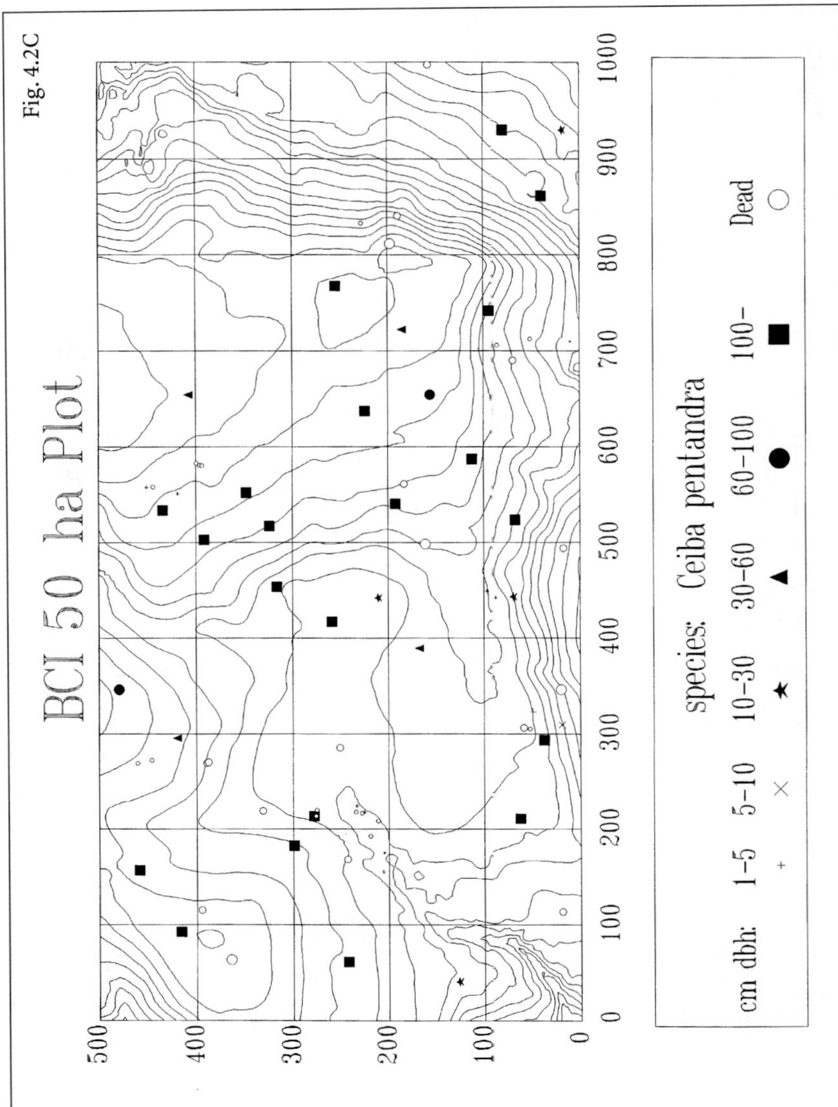
Fig. 4.2C
BCI 50 ha Plot
cm dbh: 1–5 5–10 10–30 30–60 60–100 100– Dead
species: Ceiba pentandra
0 100 200 300 400 500 600 700 800 900 1000
0 100 200 300 400 500

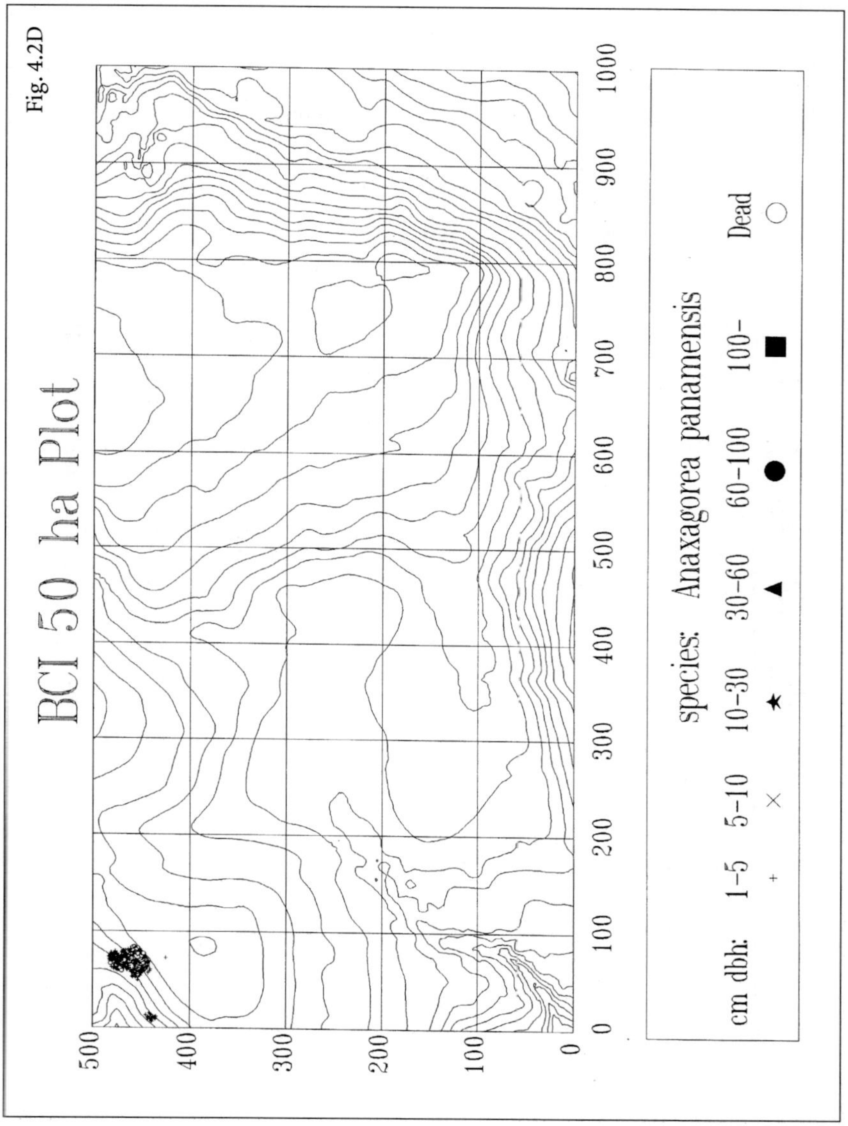
BCI 50 ha Plot
0
100
200
300
400
500
600
700
800
900
1000
cm dbh: 1–5 5–10 10–30 30–60 60–100 100– Dead
species: Anaxagorea panamensis

Fig. 4.2D

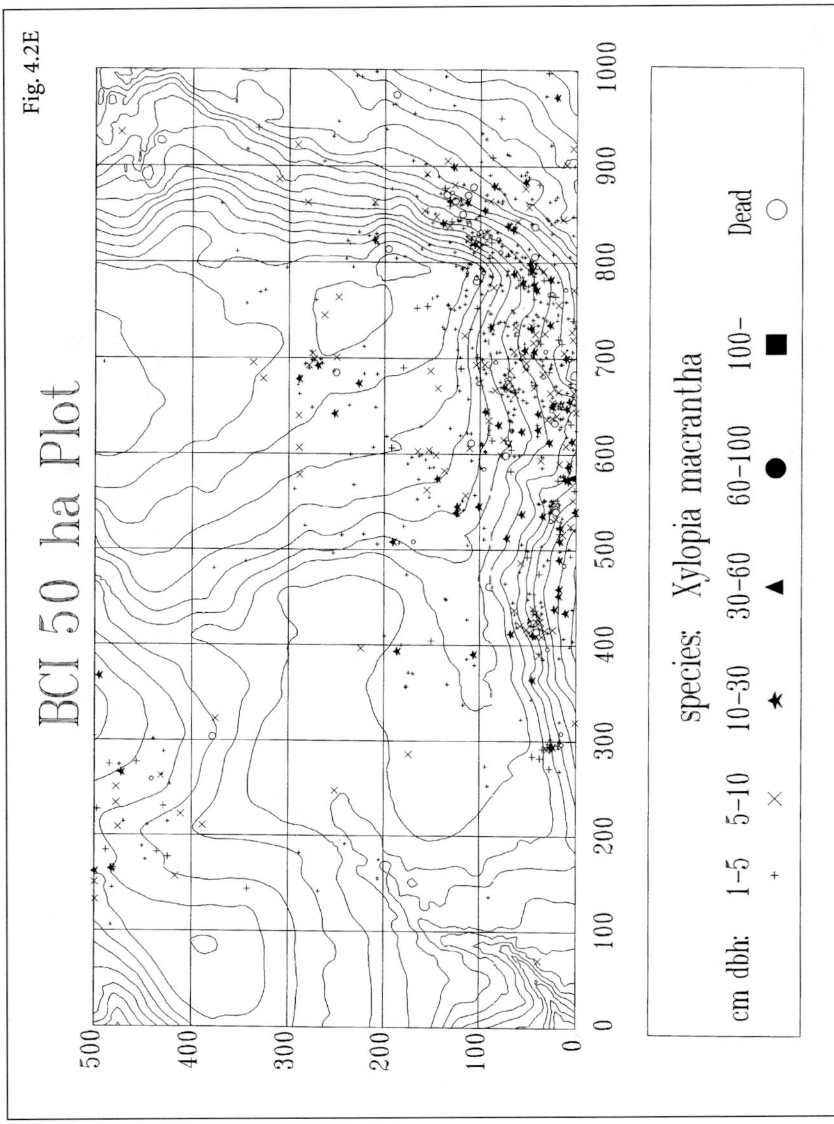
BCI 50 ha Plot
500
400
300
200
100
0
0
100
200
300
400
500
600
700
800
900
1000
species: Xylopia macrantha
cm dbh: 1–5 5–10 10–30 30–60 60–100 100– Dead

Fig. 4.2E

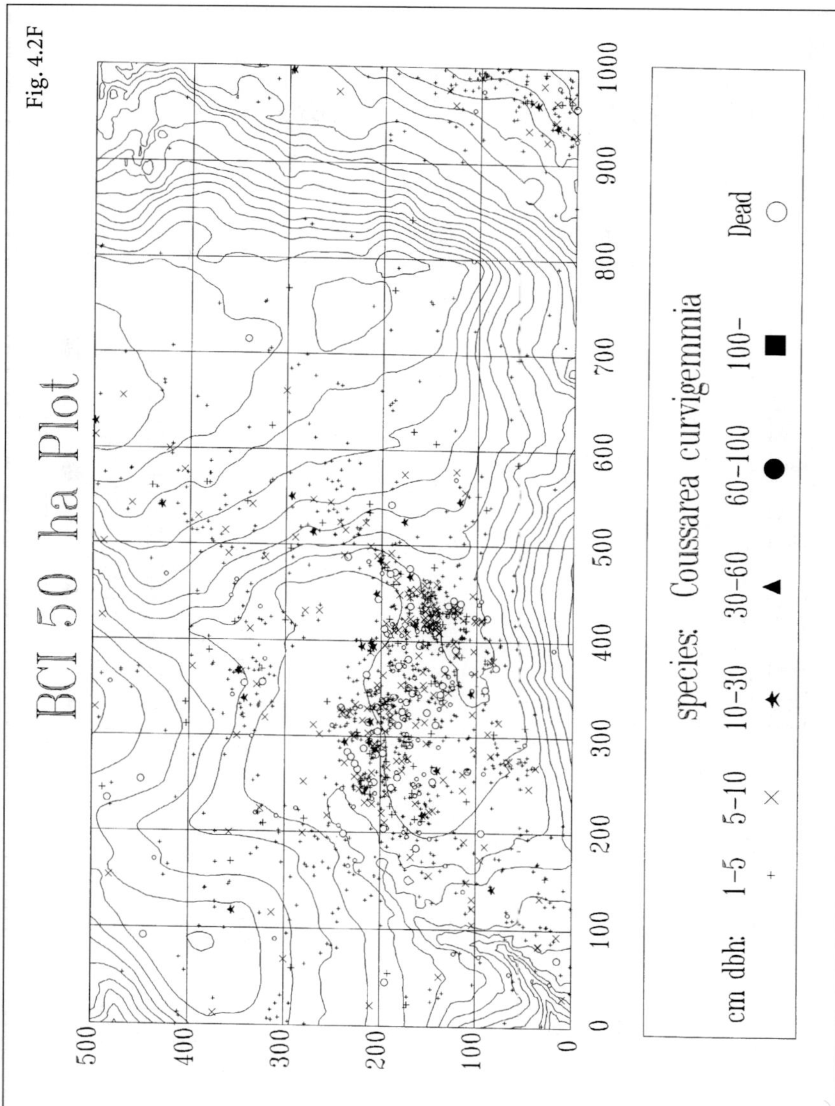
BCI 50 ha Plot
500
400
300
200
100
0
0 100 200 300 400 500 600 700 800 900 1000
cm dbh: 1–5 5–10 10–30 30–60 60–100 100– Dead
species: Coussarea curvigemmia

Fig. 4.2F

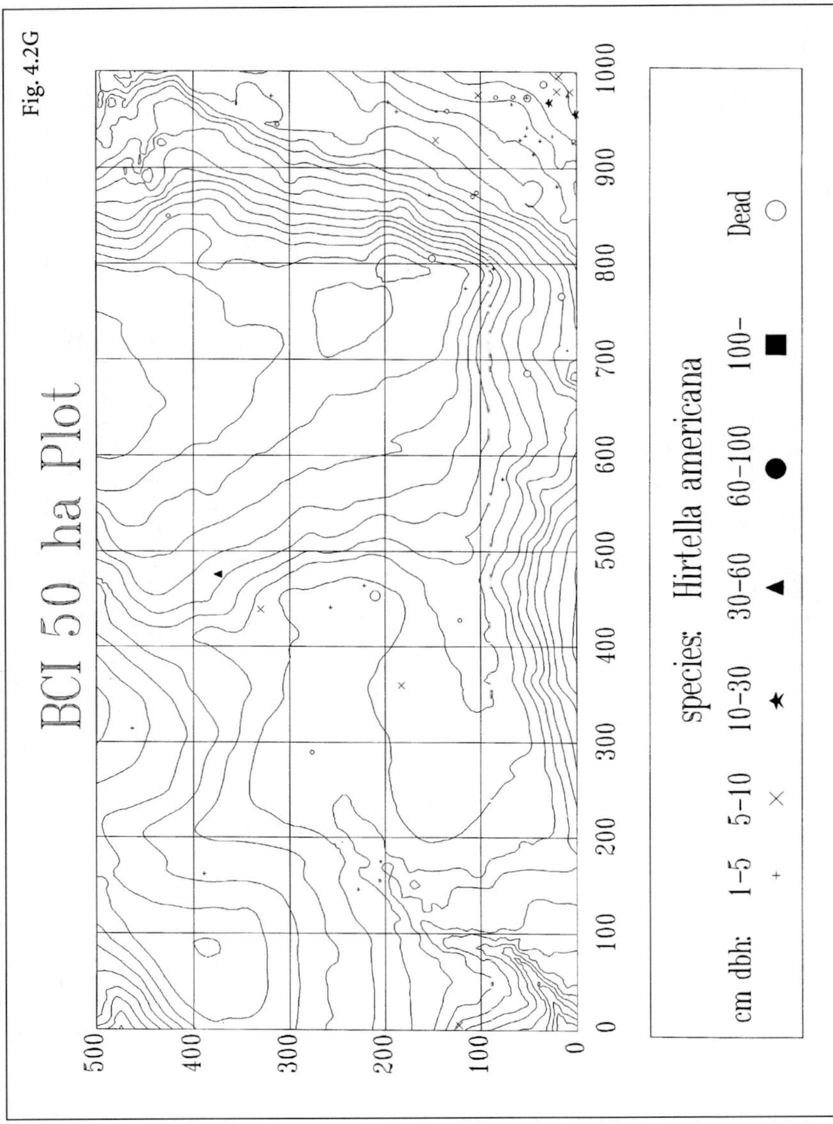
BCI 50 ha Plot
species: Hirtella americana
cm dbh: 1–5 5–10 10–30 30–60 60–100 100– Dead
0 100 200 300 400 500 600 700 800 900 1000
0 100 200 300 400 500

Fig. 4.2G

Fig. 4.2H

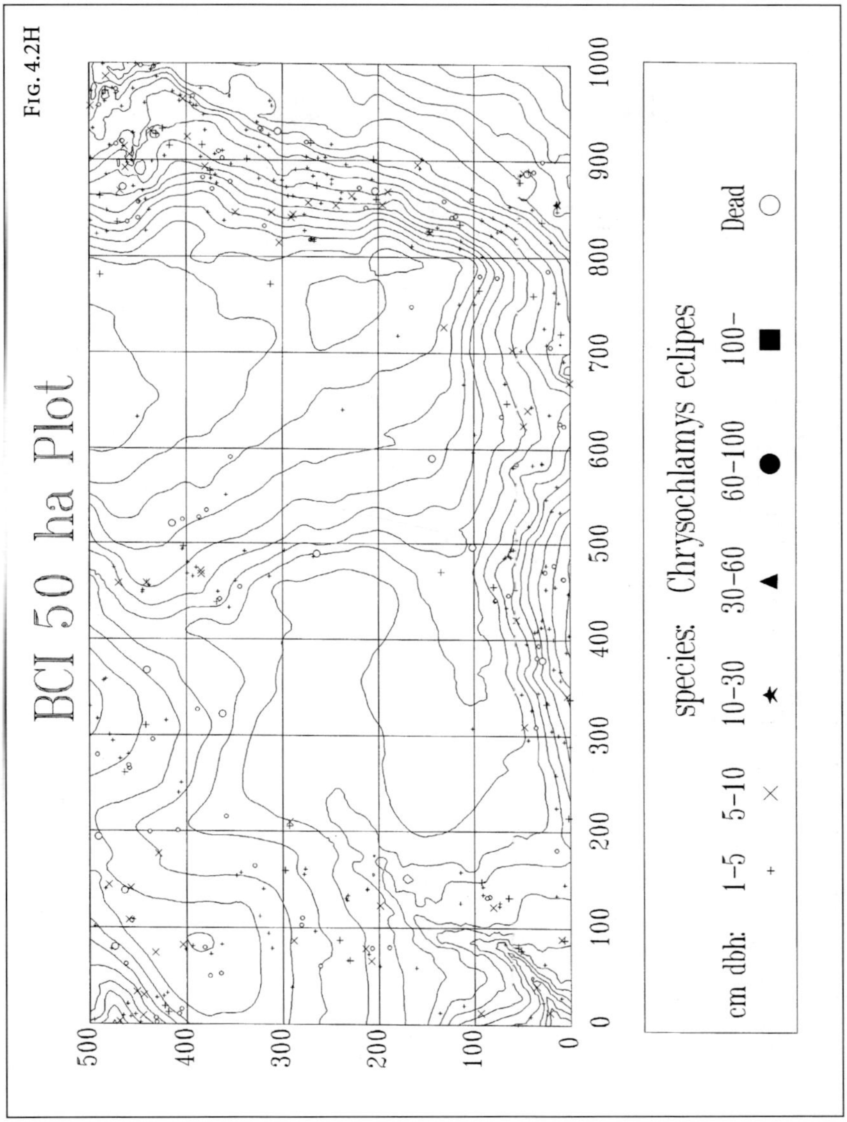
BCI 50 ha Plot
0
100
200
300
400
500
600
700
800
900
1000
species: Chrysochlamys eclipes
cm dbh: 1–5 5–10 10–30 30–60 60–100 100– Dead

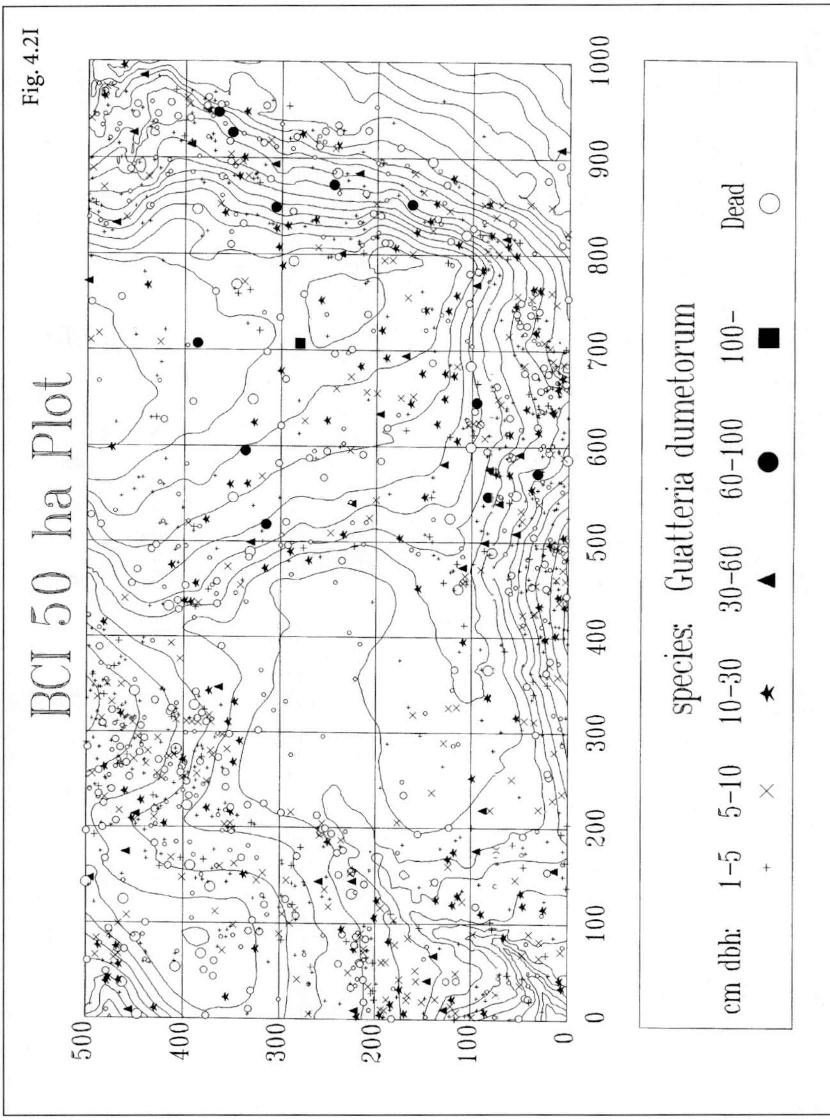
BCI 50 ha Plot
0
100
200
300
400
500
600
700
800
900
1000
0
100
200
300
400
500
species: Guatteria dumetorum
cm dbh: 1–5 5–10 10–30 30–60 60–100 100– Dead

Fig. 4.21

FIG. 4.2J

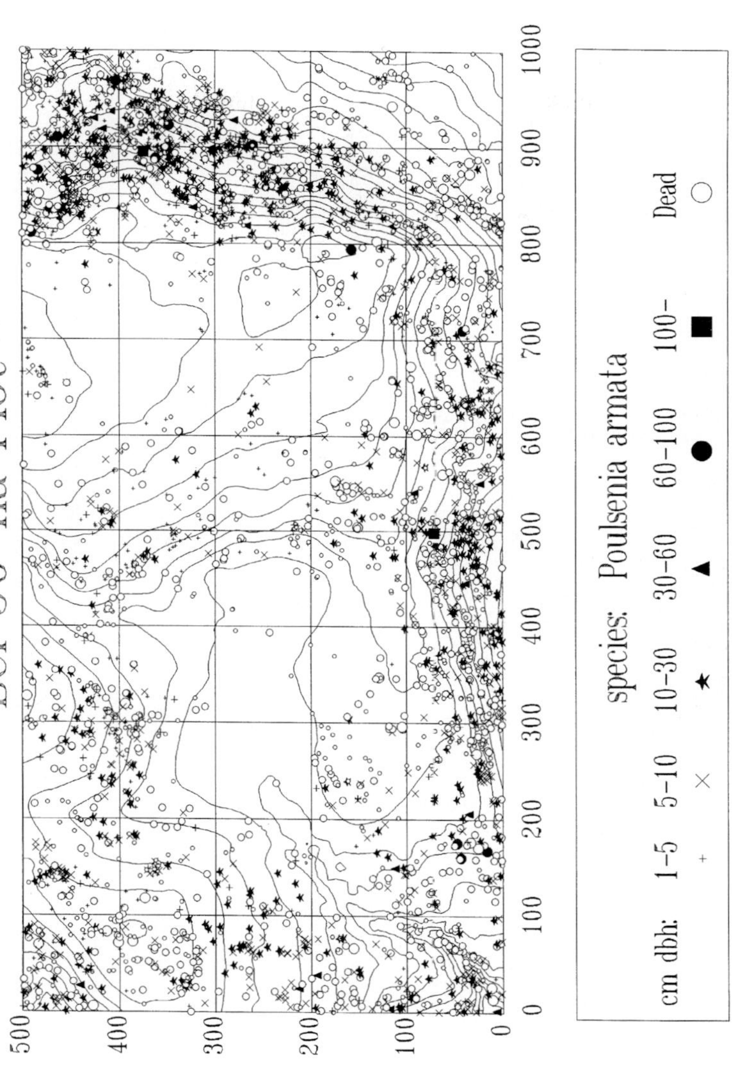
BCI 50 ha Plot
500
400
300
200
100
0
0 100 200 300 400 500 600 700 800 900 1000
species: Poulsenia armata
cm dbh: 1-5 5-10 10-30 30-60 60-100 100- Dead

Fig. 4.2K

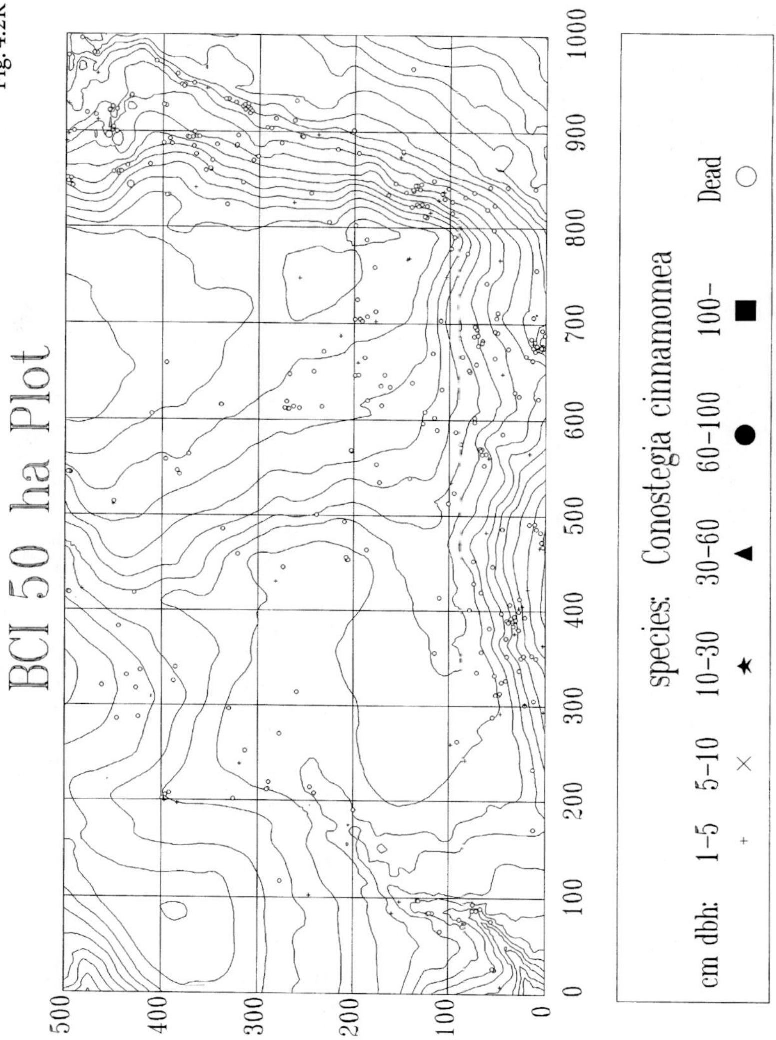
BCI 50 ha Plot
0
100
200
300
400
500
600
700
800
900
1000
species: Conostegia cinnamomea
cm dbh: 1–5 5–10 10–30 30–60 60–100 100– Dead

Fig. 4.2L

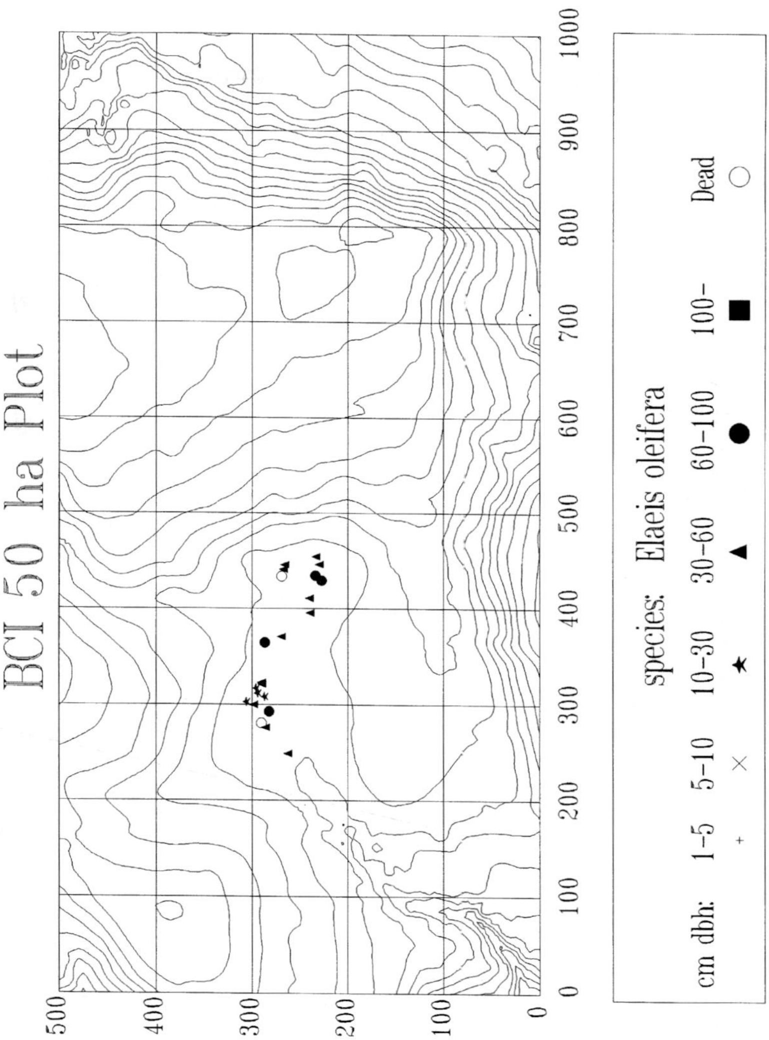
BCI 50 ha Plot
0 100 200 300 400 500
0 100 200 300 400 500 600 700 800 900 1000
species: Elaeis oleifera
cm dbh: 1–5 5–10 10–30 30–60 60–100 100– Dead

Fig. 4.2M

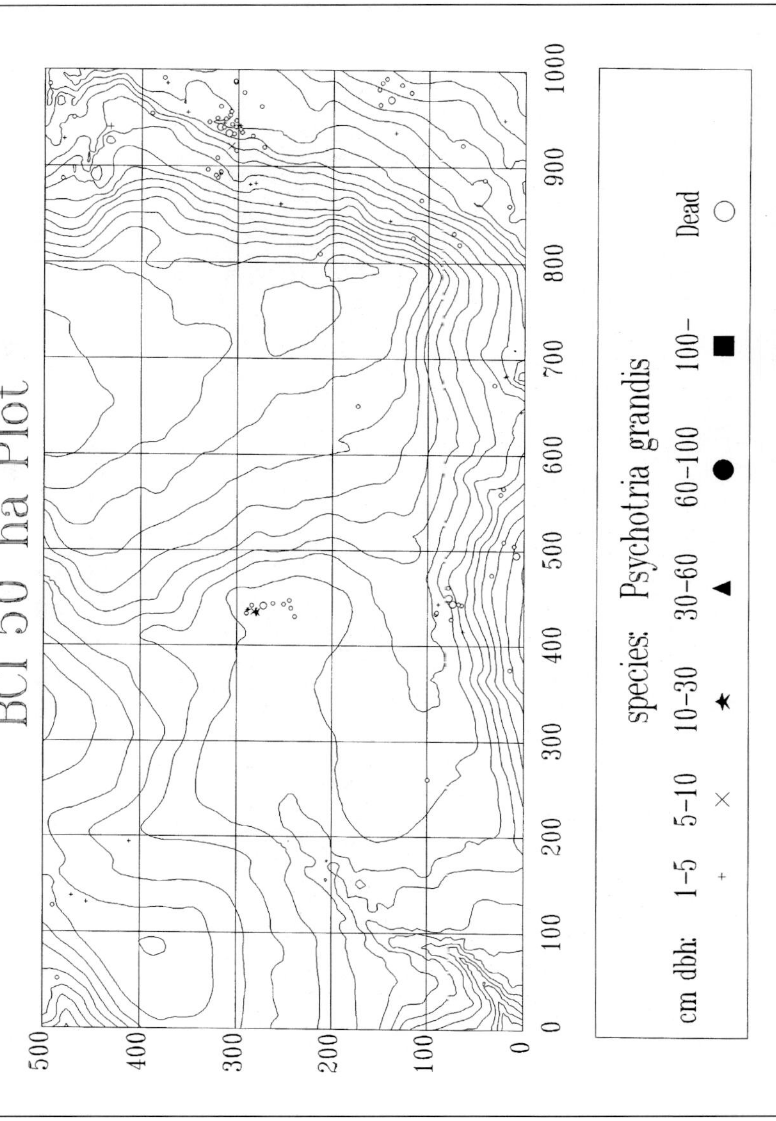
BCI 50 ha Plot
500
400
300
200
100
0
0
100
200
300
400
500
600
700
800
900
1000
species: Psychotria grandis
cm dbh: 1–5 5–10 10–30 30–60 60–100 100– Dead

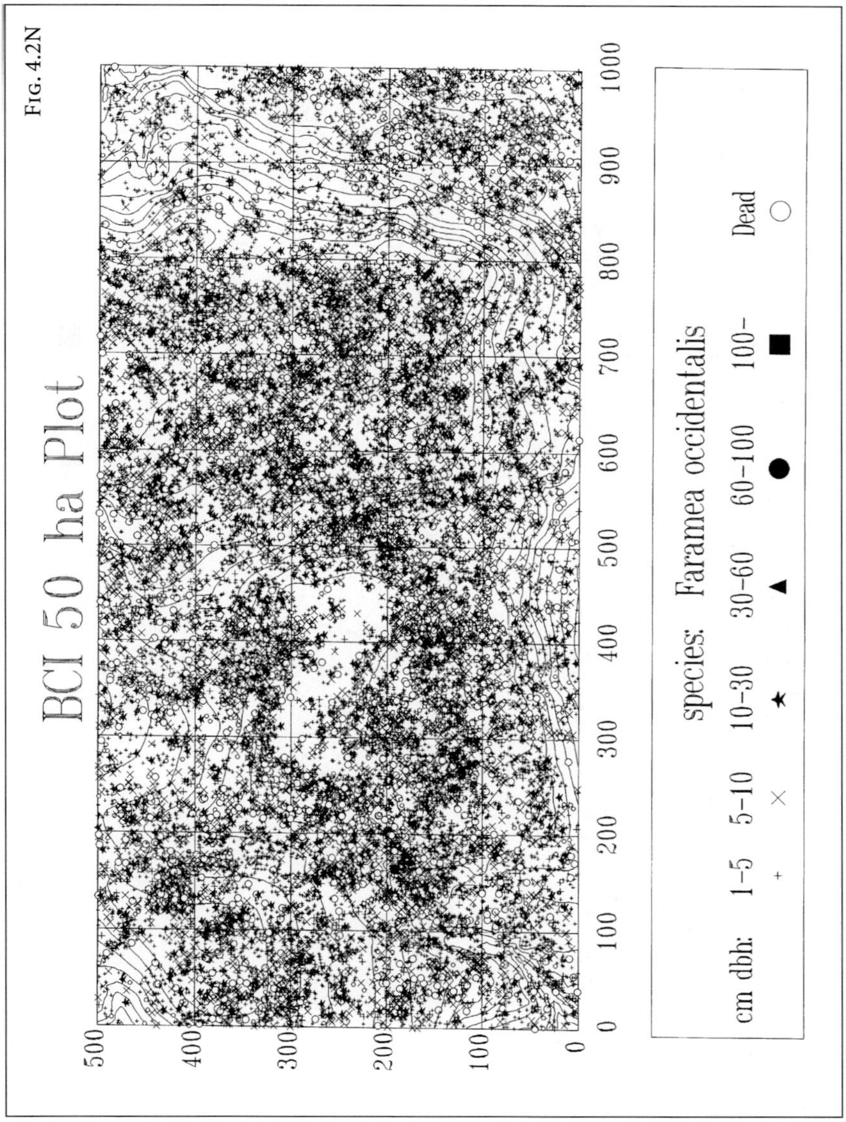
BCI 50 ha Plot
0
100
200
300
400
500
600
700
800
900
1000
species: Faramea occidentalis
cm dbh: 1–5 5–10 10–30 30–60 60–100 100– Dead

Fig. 4.2N

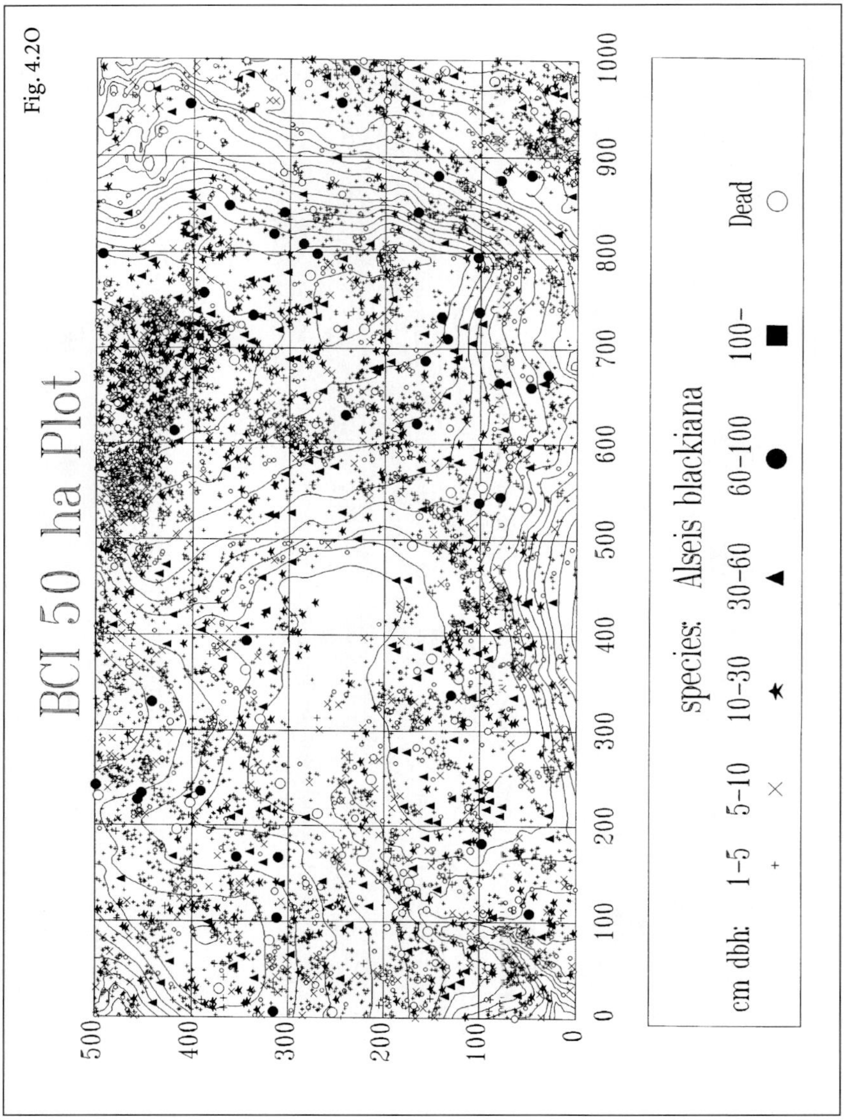
BCI 50 ha Plot
0
100
200
300
400
500
600
700
800
900
1000
species: Alseis blackiana
cm dbh: 1–5 5–10 10–30 30–60 60–100 100– Dead

Fig. 4.20

FIG. 4.2P

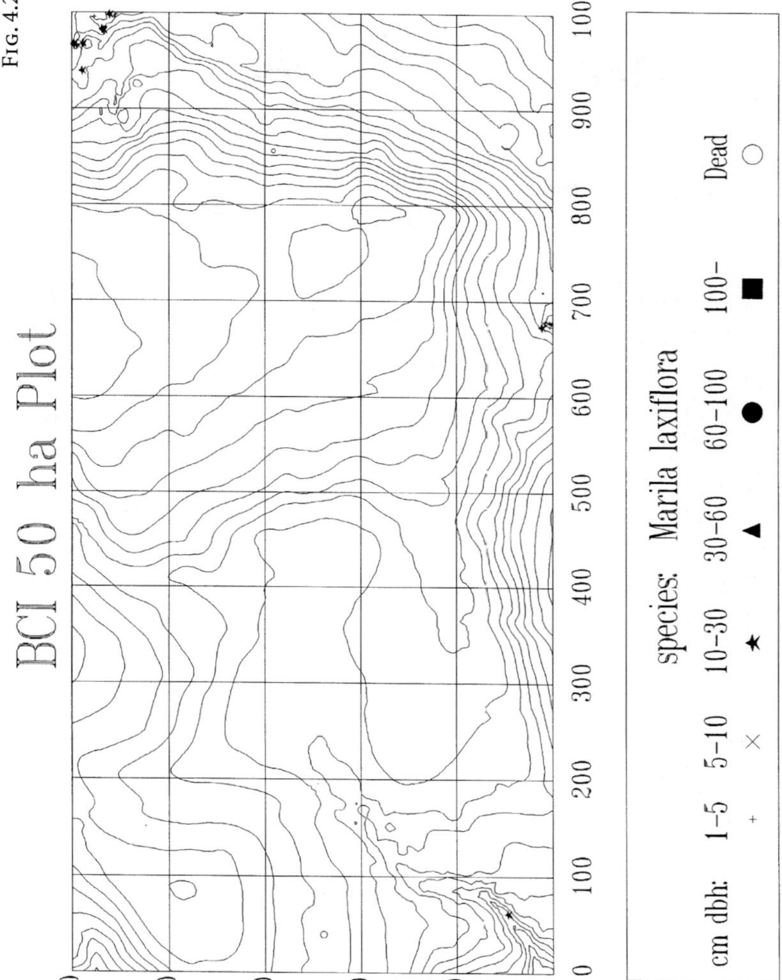
BCI 50 ha Plot
species: Marila laxiflora
cm dbh: 1–5 5–10 10–30 30–60 60–100 100– Dead
0 100 200 300 400 500 600 700 800 900 1000
0 100 200 300 400 500

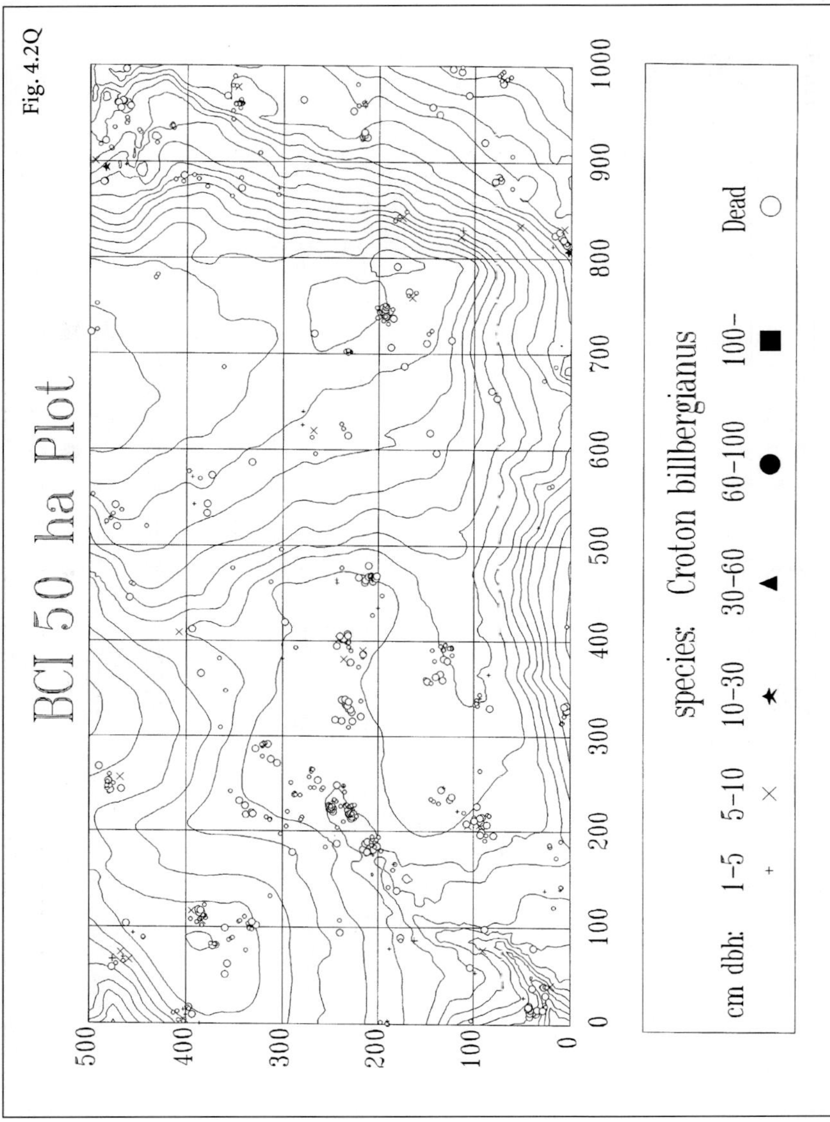
Fig. 4.2Q
BCI 50 ha Plot
species: Croton billbergianus
cm dbh: 1–5 5–10 10–30 30–60 60–100 100– Dead
0 100 200 300 400 500 600 700 800 900 1000
0 100 200 300 400 500

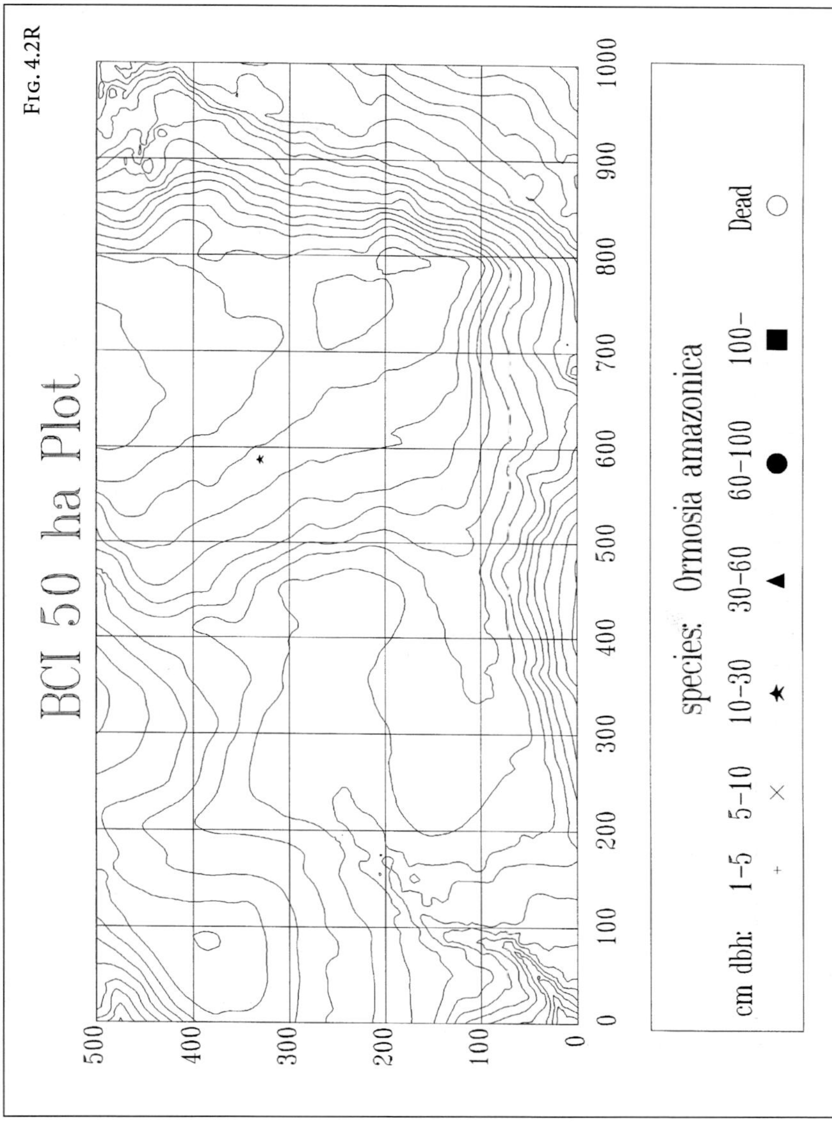
BCI 50 ha Plot
0
100
200
300
400
500
600
700
800
900
1000
species: Ormosia amazonica
cm dbh: 1–5 5–10 10–30 30–60 60–100 100– Dead

Fig. 4.2R

References

Becker P, Rabenold PE, Idol JR, Smith AP. 1988. Water potential gradients for gaps and slopes in a Panamanian tropical moist forest's dry season. Journal of Tropical Ecology, 4:173-184.

Clark DB, Clark DA. 1996. Abundance, growth, and mortality of very large trees in neotropical lowland rain forest. Forest Ecology and Management, 80:235-244.

Condit R. 1995. Research in large, long-term tropical forest plots. Trends in Ecology and Evolution, 10:18-22.

Condit R, Hubbell SP, Foster RB. 1992. Recruitment near conspecific adults and the maintenance of tree and shrub diversity in a neotropical forest. American Naturalist, 140:261-286.

Condit R, Hubbell SP, Foster RB. 1993a. Identifying fast-growing native trees from the neotropics using data from a large, permanent census plot. Forest Ecology and Management, 62:123-143.

Condit R, Hubbell SP, Foster RB. 1993b. Mortality and growth of a commercial hardwood, "El Cativo", *Prioria copaifera*, in Panama. Forest Ecology and Management, 62:107-122.

Condit R, Hubbell SP, Foster RB. 1994. Density dependence in two understory tree species in a neotropical forest. Ecology, 75:671-705.

Condit R, Hubbell SP, Foster RB. 1995a. Mortality rates of 205 neotropical tree species and the responses to a severe drought. Ecological Monographs 65:419-439.

Condit R, Hubbell SP, Foster RB. 1995b. Wood volume and production of tropical timber species: data from a large, permanent plot in Panama. Journal of Tropical Forest Science, 7:599-622.

Condit R, Hubbell SP, Foster RB. 1996a. Assessing the Response of Plant Functional Types in Tropical Forests to Climatic Change. Journal of Vegetation Science. 7:405-416.

Condit R, Hubbell SP, Foster RB. 1996b. Changes in a tropical forest with a shifting climate: results from a 50 ha permanent census plot in Panama. Journal of Tropical Ecology, 12:231-256.

Condit R, Hubbell SP, LaFrankie JV, Sukumar R, Manokaran N, Foster RB, Ashton PS. 1996c. Species-Area and Species-Individual Relationships for Tropical Trees: A Comparison of Three 50 Ha Plots. Journal of Ecology, 84:549-562.

Condit R, Foster RB, Hubbell SP, Sukumar R, Leigh EG, Manokaran N, Loo de Lao S. 1998. Assessing Forest Diversity from Small Plots: Calibration using Species-Individual Curves from 50 Ha Plots. In: Dallmeier F, ed. Measuring and Monitoring Forest Biodiversity: The International Network of Biodiversity Plots. Smithsonian Institution/Man and the Biosphere. (in press)

Dallmeier F, Kabel M, Taylor C, Romano CB, Rice R, Mayne JC. 1991. User's Guide to the Bisley Biodiversity Plots: Luquillo Biosphere Reserve - Puerto Rico. Contribution #7, Smithsonian Institution/Man and the Biosphere. Washington, D.C.

De Steven D. 1994. Tropical tree seedling dynamics: recruitment patterns and their population consequences for three canopy species in Panama. Journal of Tropical Ecology, 10:369-383.

Decker C, Bolstad PV. 1996. Forest canopy, terrain, and distance effects on global positioning system point accuracy. Photogrammetric Engineering and Remote Sensing, 62:317-321.

Elias TS. 1980. The Complete Trees of North America. Van Nostrand Reinhold Company, New York. 948 pages.

Garwood NC. 1983. Seed germination in a seasonal tropical forest in Panama: a community study. Ecological Monographs, 53:159-181.

Hall P, Ashton PS, Condit R, Manokaran N, Hubbell SP. 1998. Signal and noise in sampling tropical forest structure and dynamics. In: Dallmeier F, ed. Measuring and Monitoring Forest Biodiversity: The International Network of Biodiversity Plots. Smithsonian Institution/Man and the Biosphere. (in press)

Harms K. 1997. The maintenance of diversity in a neotropical tree community of Panama. Ph.D. dissertation. Princeton University, Princeton, NJ.

Hubbell SP. 1998. A Theory of Biogeography and Relative Species Abundance. Princeton Monographs in Population Biology, Princeton University Press, Princeton, NJ. (in press)

Hubbell SP, Foster RB. 1983. Diversity of canopy trees in a neotropical forest and implications for conservation. In: Sutton SL, Whitmore TC, Chadwick AC, eds. Tropical Rain Forest: Ecology and Management. Blackwell Scientific Publications, Oxford, pp. 25-41

Hubbell SP, Foster RB. 1986a. Biology, chance, and history and the structure of tropical rain forest tree communities. In: Diamond J, Case TJ, eds. Community Ecology. New York: Harper and Row, New York, pp. 314-329

Hubbell SP, Foster RB. 1986b. Commonness and rarity in a neotropical forest: implications for tropical tree conservation. In: Soulé M, ed. Conservation Biology: The Science of Scarcity and Diversity. Sunderland, Massachusetts: Sinauer Associates, Inc., pp. 205-231.

Hubbell SP, Foster RB. 1992. Short-term population dynamics of a neotropical forest: why ecological research matters to tropical conservation and management. Oikos, 63:48-61.

Manokaran N, LaFrankie JV, Kochummen KM, Quah ES, Klahn J, Ashton PS, Hubbell SP. 1990. Methodology for the 50-hectare research plot at Pasoh Forest Reserve. Forest Research Institute of Malaysia, Research Pamphlet No. 104, 69 pages. Kepong, Malaysia.

Manokaran N, LaFrankie JV, Kochummen KM, Quah ES, Klahn J, Ashton PS, Hubbell SP. 1992. Stand table and distribution of species in the 50-ha research plot at Pasoh Forest Reserve. Forest Research Institute of Malaysia, Research Data No. 1, 454 pages. Kepong, Malaysia.

Murawski DA, Hamrick JL, Hubbell SP, Foster RB. 1990. Mating systems of two Bombacaceous trees of a neotropical moist forest. Oecologia, 82:501-506.

Phillips OL, Gentry AH. 1994. Increasing turnover through time in tropical forests. Science, 263:954-958.

Phillips OL, Hall P, Gentry AH, Sawyer SA, Vásquez R. 1994. Dynamics and species richness of tropical rain forests. Proceedings of the National Academy of Science, 91:2805-2809.

Saw LG, LaFrankie JV, Kochummen KM, SK Yap. 1991. Fruit trees in a Malaysian rain forest. Economic Botany, 45:120-136.

Sheil D. 1995. A critique of permanent plot methods and analysis with examples from Budongo Forest, Uganda. Forest Ecology and Management, 77:11-34.

Stacy EA, Hamrick JL, Nason JD, Hubbell SP, Foster RB, Condit R. 1996. Pollen dispersal in low density populations of three neotropical tree species. American Naturalist, 148:275-298.

Swaine MD, Lieberman D, Putz FE. 1987. The dynamics of tree populations in tropical forest: a review. Journal of Tropical Ecology, 3:359-366.

Sukumar R, Dattaraja HS, Suresh HS, Radhakrishnan J, Vasudeva R, Nirmala S, Joshi NV. 1992. Long-term monitoring of vegetation in a tropical deciduous forest in Mudumalai, southern India. Current Science, 62:608-616.

Welden CW, Hewett SW, Hubbell SP, Foster RB. 1991. Survival, growth, and recruitment of saplings in canopy gaps and forest understory on Barro Colorado Island, Panamá. Ecology, 72:35-50.

Yamakura T, Kanzaki M, Itoh A, Ohkubo T, Oginok K, E Chai OK, Lee HS, Ashton PS. 1995. Topography of a large-scale research plot established within a tropical rain forest at Lambir, Sarawak. Tropics, 5:41-56.

Zimmerman JK, Everham EM III, Waide RB, Lodge DJ, Taylor CM, Brokaw NVL. 1994. Responses of tree species to hurricane winds in subtropical wet forest in Puerto Rico: implications for tropical tree life histories. Journal of Ecology, 82:911-922.

Index

A

administration 13
Africa 10, 12, 56
Amazonia 6
analyses 3-5, 12, 26, 30, 60, 62, 83, 103, 105, 106, 108, 109, 135, 137-140, 165-168
Andes 6
Araceae 39
ASCII 108, 140, 142, 146, 166, 169
Ashton 70
Asia 5, 6, 10, 12, 56, 92
assistants 13, 66, 174
Atta 58

B

Balslev 6
bamboo 39, 57
bark 46, 65-67, 80
basal area 4, 59, 61
Basic 140
basic research 9
BCI (Barro Colorado Island) 3-6, 11, 17, 23, 29, 32-34, 46, 52-62, 65, 69, 70, 73, 75, 78, 82, 86-89, 91, 92, 95, 97, 108, 112, 114, 119, 140, 164, 166, 169, 171, 173-175, 179, 181, 183
binoculars 38, 52, 62, 66, 78, 98
biologist 5, 6, 17, 37
boat 97
bole stem 60, 61
Bolstad 32
botany 37
  botanist 5, 37, 65, 70, 71, 85
  botanical 6
branch form 65
branching 66
breast height 37, 38, 47, 50, 52-54, 60, 80, 82, 83, 85, 89, 96
buttresses 50, 51, 53, 60, 66, 83, 89, 90, 107, 137, 163, 171

C

C/C++ 140
calipers 38, 46, 47, 49, 59, 62, 78, 98
  wooden 38, 47
  large 38, 47, 78
Cameroon 5-7, 11, 32, 33, 52, 56, 58, 59, 62, 97
camping gear 97
canopy 32, 57, 91-93, 182
carbon storage 9
census 3-6, 9, 11, 13, 17, 33, 37-39, 45-47, 50, 54-59, 61-63, 65, 66, 68, 70, 73-93, 95-98, 103, 105-107, 109, 112, 114, 121, 123-126, 135-142, 145, 146, 154, 159, 161-169, 171, 179
  date 55, 58, 62, 86, 106, 107, 112, 121, 124, 126, 136, 139, 142, 146, 162, 165-168
  first 11, 37, 38, 57, 59, 61, 62, 70, 77-80, 82-87, 106, 107, 109, 135, 137-139, 141, 145, 161, 163-169, 171
  interval 62, 77, 87, 140
  trees 17, 33, 37, 55, 58, 59, 61, 62, 65, 73, 77, 78, 80-83, 85, 87, 90-92, 105-107, 109, 114, 124, 135, 137, 141, 145, 154, 161, 163-167, 169
Center for Tropical Forest Science (CTFS) 3-6, 9, 10, 12, 56, 77, 173, 174, 183
checking work 56
chimpanzees 32
clerical 13, 106
clerk 113
climate 9, 10, 12
clones 54
co-author 173, 174
codes 55, 61, 66, 67, 69, 70, 80, 82-84, 86, 89, 90, 105, 107, 112, 118, 122, 124, 135-139, 141, 142, 145, 159, 161-163, 165, 167-169, 171
  big-tree 44, 47, 50-52, 55, 78, 83, 85, 108, 111, 113, 114, 137, 138, 140-142, 145, 163, 167, 171
  broken stem 83, 164
  buttressed tree 38, 51, 54, 60, 61, 89, 90, 161
  death 85, 161
  general 138, 139, 161, 163-165, 168
  irregular stem 63, 82
  leaning stem 47, 52, 60
  multiple stems 50, 53, 54, 83-86, 89, 107, 127, 142, 161, 163, 171
  prior 80, 85, 89, 159, 165
  problem 43, 50, 55, 80, 84-86
  prostrate stem 47, 52, 53, 85
  resprout 82-86, 89, 90, 95, 154, 161, 163, 164, 171
  status 138, 139, 146, 161, 163-169
collecting 56, 62, 65, 66, 68-71, 136, 137, 139, 162, 168

Colombia 6, 10, 11, 57, 99
column 23-26, 30, 39, 45, 46, 50, 51, 54, 55, 59, 67, 80, 83-86, 106, 110, 131, 137, 146, 148, 159, 160, 165
compass 19-21, 32-34, 38, 62, 78, 97, 98
  See also surveying
  bearing 19, 21, 24, 25, 33
  needle 19, 21
computers 13, 97, 98, 114, 173
  hand-held 105
Congo 5-7, 11, 21, 23, 26, 27, 30, 32-34, 56, 58, 62, 71, 75, 87, 93, 97
  Democratic Republic of 7
contour lines 28, 179, 181
coordinate system 20
coordinates 20, 45, 58, 108-110, 114, 131, 135, 138, 142, 144, 146, 154, 161, 165-167, 169, 183
  global 109, 110, 131, 144, 146, 154
  local 108-110, 131, 154
coordinators 13, 37, 90, 108, 111, 112, 115
  database 115, 135, 167
  field 12, 13, 37, 55, 78, 80, 90, 108, 111, 112, 115
cost 33, 56, 62, 97-99
Croat 70
Cyclanthaceae 39

D

Dallmeier 58
data
  administrative 106
  availability 173
  collection 4, 17, 32, 73, 114
  file map 112
  screening 88, 108, 109, 122, 127
data entry 12, 13, 86, 91, 98, 105, 106, 108, 112, 114, 117, 118, 121-126, 129, 135, 136, 138, 145, 162, 167
  clerk 106, 107, 108, 111-115, 121-124, 135, 142
  double 107, 108, 113-115, 121, 126, 129
  programs 103, 105-108, 112, 115, 118, 121, 133
data file 105, 106, 108, 109, 112, 129, 133, 144, 154
  main 105, 112
data sheets 12, 39, 40, 51, 55, 56, 62, 70, 78, 84-86, 89, 105-112, 114, 135, 141, 156
  main 39, 109
  map 39, 58, 87, 108, 109, 111, 112, 114, 169
  multiple-stem 39, 42, 52, 56, 62, 78, 83-85, 105, 141
  original trees 80, 81, 85, 105, 107, 114, 123, 124, 141, 161, 164, 165
  problems 4, 55, 103, 112
  recruits 105, 107, 141, 161, 163, 165, 169
database 4-6, 13, 55, 57-59, 62, 66-69, 71, 75, 78, 86-89, 103, 105-109, 111-115, 121, 122, 124-126, 130, 131, 133, 135-146, 148, 154, 155, 159, 161-169, 171, 173-175
  big tree 111
  census 78, 114, 138, 139, 141, 154, 161, 164-169
  FoxPro 103, 105-109, 111, 112, 114, 122-124, 129-131, 133, 137, 138, 140-142, 144-147, 149, 153-155, 159, 161, 164, 167, 169
  main 105-109, 111-115, 135, 137, 138, 140-145, 148, 154, 161, 163, 167
  merging 54, 109, 111, 112, 114, 154
  methods 4, 6, 32, 56, 60, 77, 103, 140
  multiple-stem 111
  permanent 139, 161, 164-166, 168
  quadrat 105, 106, 109, 112, 130, 135, 138-140, 142, 149, 161, 165, 168
  taxonomic 66, 68, 111, 135
dbh 37, 38, 46-48, 50-52, 54, 55, 59-61, 63, 73-76, 80-90, 93, 95, 96, 98, 105, 107, 109, 112, 118, 122, 124-127, 135-146, 154, 159, 161-169, 171
deciduous 68, 82, 92
demographic data 3, 9, 10, 61, 77, 91, 93, 103, 135
density 9, 46, 91-93, 95, 97, 155, 181, 182
density-dependence 9
diameter 3, 4, 9, 19, 20, 21, 23, 37, 38, 45-47, 50-53, 57, 60, 62, 78, 83, 85, 93, 171, 183
dispersal 91
distribution 4, 5, 9, 13, 19, 179, 181-183
disturbance 9, 10, 12
diversity 3, 4, 10, 93
dry season 11, 12, 179, 181
duplicate 111
  records 133
  tag numbers 56, 133
  tags 111

E

ecologist 13, 17, 37
economic 9, 174
Ecuador 5-7, 10, 11, 33, 56, 58, 62, 70, 71, 75, 95, 97
elevation 19, 21, 22, 26-31
epiphytes 47

errors 4, 17, 21, 23, 32, 33, 73, 75-78, 83, 84, 86-90, 95, 103, 105, 107-109, 111-115, 129, 140, 141, 167

F

fecundity 91
field assistants 13
flagging 20, 23, 32, 78, 97, 98
flora 6, 12, 13, 57, 65, 66, 68-71, 78, 84, 91, 109, 111, 182
forest 3-6, 9-13, 17, 19-21, 30, 32, 33, 37, 38, 57, 60, 65, 70, 91-93, 95, 98
Forest Research Institute of Malaysia 4, 6
forest-dynamics plot 10, 12
Forestry Supplier 19, 20, 38, 46, 58, 62
FORTRAN 140
FoxPro *see database*

G

GPS 32
grid 5, 19, 23, 27-30, 32-34, 37, 45, 77, 91, 95-98
grid marker 20, 21, 38, 39
growth 4, 30, 39, 46, 47, 59-62, 73, 86-89, 91, 92, 95, 135, 164, 166, 167, 171, 173

H

height-of-measure 80, 85
herbaceous plants 63
herbarium 12, 67, 71
herbs 57
host tree 54
host-country 12, 13
Hubbell 70, 92

I

identification 13, 17, 39, 55, 62, 63, 65, 66, 68-71, 75, 78, 80, 84-86, 91, 92, 97, 98, 167
inclination 19-22, 26, 30, 34, 179
India 3, 5, 10, 11, 32, 88, 138
intellectual property 103, 173, 174
Iomega 141
Ituri 5, 11, 26-28, 31-34, 56, 58, 60, 61, 71, 87, 93, 95, 96

J

juveniles 9, 57, 65, 67, 68, 181

L

La Planada 11, 29, 99
labor 12, 13, 17, 24, 33, 37, 58, 71, 76, 95, 97-99
LaFrankie 5, 6, 70
Lambir National Park 3
Lao 6
Latin America 10, 12
leaf sheath 54
leaves 51, 54, 65, 66, 68, 81, 82, 92
lianas 5, 11, 20, 39, 47, 56, 57, 68, 92, 93, 96
Liengola Bauma 6
logging 10,12
Losos 6
Luquillo Forest 3

M

Magaard 6
Makana 6
Malaysia 3-6, 10, 11, 56, 75
managers 6, 12, 13, 60, 108, 114, 141, 142, 163, 173-175
Manokaran 3-5, 19, 33, 34, 39, 58-60, 70, 71, 75, 95
map 105
Marantaceae 39
markers 19, 20, 24, 32, 33, 62, 97
Mars 9
measuring 17, 32, 37, 38, 47, 49, 52-54, 56, 58, 60, 82, 84, 88, 90, 92, 95
monocots 54, 57
montane 6, 10, 57
morphospecies 65, 67-69
mortality 4, 61, 62, 69, 87, 88, 92, 95, 135, 166, 173, 182
Mudumalai Game Reserve 3
multiple-stemmed plants 46, 54, 55, 61, 63, 90, 95, 137, 161, 171
Musaceae 57

N

Ndomba 6
network 4-6, 9-12, 77, 87, 96, 173, 174

O

office 12, 13, 56, 58

P

Palanan 6, 11, 96
palms 38, 54, 57, 63, 84, 95, 96
Panama 3, 6, 7, 10, 11, 70, 95, 174, 175
Pascal 140
Pasoh Forest Reserve 3
Peradeniya University 30
personnel 9, 13, 107, 138
petioles 54
Philippines 7, 10, 11
photograph, hemispheric 91
plantation 9
plants 39, 56, 57, 78, 166
plot 1, 3-6, 9, 10, 12, 13, 17, 19-21, 23-35, 37-39, 46, 56-63, 65, 66, 69-71, 75, 77, 80, 85-93, 95, 97, 99, 179, 181-183
point of measure (POM) 47, 50-52, 54, 57, 60, 61, 63, 80, 82-85, 87-90, 137, 138, 142, 146, 161, 163-169, 171
principal investigator 13, 173-175
Puerto Rico 3, 5, 7, 11, 62

Q

quadrat 20, 21, 24, 25, 30, 32-34, 38, 39, 45, 46, 55, 56, 58, 59, 62, 63, 66-68, 73, 75, 78-80, 84-86, 88-90
  numbering 39

R

rainfall 11
random samples 73, 75
range-finder 32, 92
re-identify 76, 140
re-measure 55, 86, 89, 112, 140
  re-measured 52, 63, 73, 75, 86, 106, 169
  re-measurements 50, 112
  re-measuring 30, 73, 76, 87, 88
re-measured 167
re-sampling 73
recensus 17, 52, 61, 75, 77-80, 82-90, 96-98, 105-109, 111, 113, 114, 117, 123, 124, 131, 138, 141, 149, 153, 156, 159, 161, 163-165, 167, 171
recruit 77-80, 82-88, 90, 106-109, 111, 113, 117, 161, 162, 164, 165, 169, 182
resprout 96
rooting point 47, 52, 56
roots 46, 47, 53, 66, 88, 95
row 23-26, 30, 34, 110, 131, 148, 159, 160

S

sap 65
saplings 9, 20, 37, 59, 66, 73, 77, 81, 91, 92
savanna 12
seed trap 93
seedling 91-93
seeds 91-93
shrubs 10, 37, 91
sighting 19-21, 23, 26-30, 33, 34
Sigma Scan 108
Sinharaja 5, 6, 11, 29, 30, 34, 56, 57, 60, 71, 93, 96
site selection 10
slope 19-22, 24, 26, 179, 182
Smithsonian Tropical Research Institute 6
soil 10, 12
species 3-5, 9-11, 13
specimens 12, 65-71, 97
sprout 53, 60, 81-83, 88, 90
Spyglass 28, 29, 31
Sri Lanka 5, 7, 10, 11, 56, 62, 71, 93, 97
stakes 19-27, 29, 30, 32-34, 38, 39, 58, 62, 77, 78, 97, 98
  aluminum 32
  plastic tent 33
  PVC 20, 62, 98
  steel 32
standard 4, 24, 30, 32, 52, 56, 58, 62, 87, 91, 122
  standardized 4, 10, 13, 17, 30, 63, 77, 95, 174
stem 3, 4, 9, 23, 37-39, 46-48, 50, 52-61, 63, 70, 73, 75, 77, 78, 82-90, 93, 95, 96, 107, 118, 126, 127, 137, 138, 142-144, 154, 160, 167, 171
sterile 6, 65, 70
stipules 65
stranglers 39, 47, 49, 54, 57, 63, 84, 95, 96
stream 24, 34, 58, 179, 181, 182
string 46, 58, 62, 80, 97, 98, 117, 139, 162, 168
subquadrat 25, 38, 39, 45, 46, 55, 57, 58, 67, 73, 80, 82, 84, 86, 88, 91, 105, 107, 109, 113, 122, 124, 126, 135, 136, 138, 139, 141, 142, 155, 159, 160-162, 165, 167-169
supervisor 12, 13, 37, 38, 54-56, 73, 75, 78, 80, 81, 84-86
surveying 13, 17, 19-21, 23, 24, 27, 29-34, 78, 97, 98
  survey 20, 21, 24, 26, 29, 30, 32-34, 75
    See also compass: surveying compass
surveyor 20, 21, 29, 30, 33
sustainable 3, 9
Sygraph 103, 141, 155

T

tag 39, 46, 52, 54-56, 59, 61-63, 66-68, 77, 78, 80-88, 90, 97, 98, 105, 107, 108, 109, 111, 113, 114, 118, 120-127, 133, 135, 136, 138-142, 144-146, 154, 159-163, 165-169
tag punching machine 38, 98
tape measure 19, 20, 24, 34, 47, 49, 52, 78
taxonomy 17, 55, 56, 65-67, 69-71, 75, 78, 86, 106, 109, 111, 112, 114, 115, 135
Thailand 3, 6, 7, 10, 11
theft, risk of 33
topographic data 5, 19, 24, 26, 28, 31, 35, 71, 97, 98, 179
trainees 65
tree ferns 38, 54
tree-climber 66
trees 142
tripod 19
trunk 21, 23, 38, 45-47, 49-54, 57-60, 63, 66-68, 80, 82, 84, 88, 96
typhoon 10

U

United States 7, 10

V

value 3, 9, 20, 27, 107, 122, 123, 129, 138, 163-165, 167
vernier scale 21
vertebrate 10

W

waterproof materials 38, 49, 62, 66, 98
wire 59
  copper 59

Z

Zip drive 141

Printing and binding: Druckerei Triltsch, Würzburg